Elma Daniela Bezerra Lima
José Roberto L. de Mattos

Ethnomathematics and Law 10.639/03 in the Quilombola Community of Curiaú

Elma Daniela Bezerra Lima
José Roberto L. de Mattos

Ethnomathematics and Law 10.639/03 in the Quilombola Community of Curiaú

Teaching and learning mathematics at school through Afro-Brazilian culture

ScienciaScripts

Imprint

Cover image: www.ingimage.com

This book is a translation from the original published under ISBN 978-3-330-75641-0.

Publisher:
Sciencia Scripts
is a trademark of
Dodo Books Indian Ocean Ltd. and OmniScriptum S.R.L publishing group

120 High Road, East Finchley, London, N2 9ED, United Kingdom
Str. Armeneasca 28/1, office 1, Chisinau MD-2012, Republic of Moldova, Europe
Printed at: see last page
ISBN: 978-620-8-27795-6

SUMMARY

INTRODUCTION

According to Vygotsky (1991), learning is a process of acquiring skills, attitudes, information and values that take place in the subject's contact with reality, the environment and other people. Vygotsky emphasizes socio-historical processes, basing the learning process on the interdependence of the subjects involved in this process, that is, we all only develop when we interact with others, considering the importance of social interaction for our personal growth and individual development, it is necessary that the act of educating, when carried out in the social environment, is experienced, shared and that the pedagogical environment is thought out and organized from the socio-cultural context of the subjects involved in this process.

Thus, the construction of mathematical knowledge, when we use knowledge that is part of the students' daily lives, arises in a more spontaneous and natural way, conditioned to the students' reality, where students can be challenged daily to discover concepts of magnitude, position, location, temporality, capacity, mass, classification, quantity, number and other concepts or knowledge that can manifest themselves in the most diverse relationships that subjects establish with knowledge. In these relationships, mathematical content can be presented to students in such a way that they can understand it based on previous knowledge, learned in contact with the natural and social environment, serving as a starting point for the construction of knowledge in the classroom.

In March 2003, the federal government sanctioned Law No. 10.639/03-MEC (BRASIL, 2003), which amends the Law of Guidelines and Bases (LDB), establishing the Curricular Guidelines for its implementation, including the compulsory subject of "Afro-Brazilian History and Culture" in the official school curriculum, in order to recover the historical contribution of black people to the construction and shaping of Brazilian society. On March 21, 2003, it created the Special Secretariat for Policies to Promote Racial Equality - SEPPIR - and instituted the National Policy to Promote Racial Equality, put the racial issue back on the national agenda and the importance of adopting affirmative public policies that positively promote transformations in the reality experienced by the black population that lead to the construction of a truly democratic, just and egalitarian society, with the aim of reversing the damage and perverse effects caused by centuries of prejudice, discrimination, racism and ignorance. It is thus taking on the historic commitment to combat racism and promote equal opportunities for the different ethnic groups that make up the Brazilian nation, ensuring the right to equal living conditions and citizenship, guaranteeing equal rights to histories and cultures, and to all Brazilians. In addition to the right of access to the different sources of national culture, which value the history and culture of Afro-Brazilians and Africans, proposing the dissemination and production of knowledge, the formation of attitudes, postures and values that educate citizens proud of their ethnic-racial belonging, descendants of Africans, who interact in the construction of a democratic nation, in which everyone, equally, has their rights guaranteed and their identity valued.

When talking about Law No. 10.639/03, Trigo states that "the debate on education at the service of diversity becomes imperative, with the challenge of affirming and revitalizing the self-image of black people" (TRIGO, 2013, p. 36), stressing that there is no privileged cultural place from which to judge other cultures, because as Paulo Freire said - "There is no knowing more or knowing less: there is different knowledge" (FREIRE, 1987, p. 68). Still for Trigo (2013), the most appropriate act of educating or teaching is one in which the intention is to transform, rescuing the value of secularly discriminated cultures, deconstructing stereotypes linked to Africa and Africans of African descent, teaching diversity in the school environment can help to deconstruct these prejudices, because Brazil is a country that was born out of the meeting of the cultures of different civilizations, because when the Europeans arrived in Brazil they found the indigenous population (owners of the land), then came the Africans; and the last wave of migrants was the Asians. The meeting of these different peoples formed our roots, which are part of our formation as Brazilian citizens.

For Trigo - the approach to African themes needs to move away from folklore and emphasize the real contributions of the African matrix" (TRIGO, 2013, p.37). The author suggests that we must first get to know Africa, which is unknown to many, so that we can deconstruct the idea that Africa is a unit and that Africans are all the same. We must get to know the human and cultural diversity of the African continent, with its more than 800 million inhabitants, spread over 54 independent countries, with more than 2,000 different languages and many ethnic groups, where the African people live together with different beliefs and cultures. We must recognize that this cultural diversity also existed at the time the Africans were brought to Brazil, because different peoples came to Brazil, with different languages and cultures, promoting the breaking down of prejudices and paradigms, facing head-on the problems relating to the ethnic-racial issue in Brazil.

The author believes that it is important that the history of Africa "is worked on from a positive perspective, not limited to denouncing the misery that affects a significant portion of its population" (TRIGO, 2013, P.37).), highlighting the preservation of the historical memory of ancient civilizations such as the Nubians and the Egyptians, who contributed greatly to the development of humanity, as well as the historical and cultural past of pre-colonial political organizations such as the kingdoms of Congo, Mali and Zambabwe, with the aim of removing black people from the position of victim and placing them as historical subjects, capable of resisting and imposing the force of their culture. Not only to show how Africans and Afro-descendants were represented culturally, but also to highlight the importance of these cultural productions where there was a black presence, especially in the teaching of Mathematics, despite the apparent difficulty of addressing the ethnic-racial theme in this discipline, in order to deconstruct the idea that the African people contributed little to the development of scientific knowledge, since we all know and recognize that Africa was the cradle of humanity and, consequently, of human knowledge.

We must show that African peoples created tools, techniques and work systems that contributed, for example, to the development of agricultural production and mineral exploration. There is a lot of African knowledge that we teachers can bring to the classroom, where we can work on the identity and contributions of the African people, recognizing and valuing the participation of black people in the construction of national culture, analyzing situations of racial diversity in the daily lives of our students. And critically analyze facts and acts of discrimination and racism, breaking with preconceived ideas about the participation of black people in the construction of knowledge, so that the racial issue becomes real and transformative knowledge.

Given all this, we asked ourselves: How do the math teachers at the school in the Quilombola community of Curiaù develop their work with their students, bearing in mind Federal Law 10.639/03? How does this law relate to the content of math classes? What activities are carried out at the Curiaù Quilombola Community school in the area of Ethnomathematics? Is there any work on flour production in this community in the area of Mathematics?

These questions disturbed us and led us to research the teaching of mathematics in the school of the Quilombola community of Curiaù, which gave us the opportunity to write this book, awakening our interest in the subject and in everything related to the African continent.

This book is divided as follows:

First of all, we present the Quilombola community of Curiaù. A brief history of the Curiaù Environmental Preservation Area - APA, in the state of Amapà, 8 km from the city of Macapà, created by state decree 024 in 1990, with the aim of protecting and conserving the natural and environmental resources of the site, which was inhabited by communities formed by slaves brought to Amapà in the 18th century. Today, around 1,500 people live in the Quilombola community of Curiaù, which is considered to be a historical and ecological site whose main economic activity is subsistence farming and plant and animal extraction.

Secondly, we highlight the current situation in Mathematics Education, what the authors point out to us as the problems that create learning difficulties for students. We also highlight Law 10.639/03 and the ways in which Ethnomathematics can help us overcome some of these problems, if we see them as challenges. Considering that the reality in which we are inserted is where knowledge first emerges, and that if we conduct our pedagogical work from the perspective of Ethnomathematics we can provide our students with experiences that will make sense when inserted within the cultural, social and natural environment in which they live.

Next, we present the school located in the community of Curiàu. We present the way in which Mathematics content is taught by the school's teachers. We describe how the teachers use the guidelines of Law 10.639/03 during the activities carried out in math classes.

Below, we present an activity carried out at the school in the Quilombola community of Curiaù, where we describe in detail how a workshop was held using the game Mancala.

Finally, we described the process of flour production and the work of the farmers in the Curiaù community, emphasizing the presence of Ethnomathematics in the work of the community's flour producers, where it was possible to observe that "ethnomathematical" principles are always present in the entire context of the research. We tried to analyze the application of mathematics content in the pedagogical practice of the mathematics teachers at the Curiàu school and the workers who produce flour and who live in the community.

We hope that this book will contribute to improving the quality of mathematics teaching and also that it will encourage new work, giving continuity to this book, which still has a lot to be explored.

THE CURIAÙ

Marin (1997) reports that the origins of Curiaù are wrapped up in a mythical tradition, built up collectively and based on facts told by residents concerned with preserving the memory of their ancestors. The quilombola community of Curiâu, according to Marin (1997), is made up of remnants of the Curiaù Quilombo, and the residents are predominantly of a collective interest, with internal rules of help, cooperation and socialization. Social customs, degrees of kinship and marriages within the community express material and symbolic exchanges that allow the group to maintain its ethnic identity, expressing feelings of belonging among those born in Curiâu, called "Criauenses" in Videira (2013), who argues that the Curiaù community is a place of memory and cultural heritage.

Marin (1997) reports that the formation of the Curiaù quilombo appears in narratives transcribed in documents from the 18th and 19th centuries in the public archives of the state of Parà and in archival pieces in the archives of French Guiana. The 1760s saw the arrival of a large number of slaves for work on the Macapà Fortress, the construction of the towns of Macapà and Mazagao and the formation of commercial agriculture.

According to Videira (2013), Curiaù is located 8 km from Macapà (see Figure 1), between parallels 00o 00' N and 00o 15'N, cut by the meridian 51° 00'W, and is bordered to the east by the Amazon River, to the north/northeast by the Pescada Igarapé and the branch that connects the EAP-070 to the BR-210, to the west by the Amapà railroad and to the south by a dry line at latitude 00o 06'N. Curiâu can be accessed by land (via the BR-210 and EAP-070) or by river (via the Curiâu River, which cuts through the Curiaù APA in an east/west direction).

In Sudam (1984), we find that the climate of the Curiaù APA is tropical-humid, with a high annual rainfall rate and a small annual temperature range, with an average annual temperature of 27°C, a maximum of 31°C and a minimum of 23°C. The average annual rainfall is 2,500 mm, the driest quarter is September, October and November, and the wettest period is March, April and May. The annual relative humidity is around 85% and the average annual sunshine is 2,200 hours. The prevailing winds come from the northern hemisphere and blow in a northeasterly direction.

Figure 1: Location of Curiaù in Amapà.

Source: Google Maps.

Facundes and Gibson (2000) explain that Curiaù has three types of soil: Yellow Latosol, Hydromorphic Soils and Alluvial Soils. The authors explain that the Yellow Latosol represents 44.22% of Curiaù's land, corresponding to 9,834.28 hectares of its surface, with flat and gently undulating relief. These are well-drained, non-hydromorphic mineral soils, with high iron content, deep, acidic, friable, with a textural class of medium to very clayey, this type of soil has low natural fertility and restrictions on agricultural practices.

According to the aforementioned authors, Hyphromorphic Soils represent 43.47% of Curiaù's surface, due to the concentration of clay and silt, this type of soil impedes vertical drainage, has good fertility, and is suitable for rice cultivation and natural pastures. Alluvial soils account for 12.31% of Curiaù's land. They are mineral soils formed by alluvial sediments, hydromorphic, eutrophic, fertile and with a flat topography.

Facundes and Gibson (2000) report that the geological formation corresponds to the Cenozoic Era, dating back approximately 65 million years, defined from the Tertiary and Quaternary periods. As for the Curiaù River, the hydrographic basin is 584.47 km long^2 , along which it runs through areas of flooded grassland covering 4.5 km within the varzea forest, until it flows into the Amazon River. The Curiaù's vegetation is made up of three predominant ecosystems: Cerrado, Flooded Fields and Varzea Forest. There are also areas of Gallery Forest, Forest Islands and Permanent Lakes.

Chagas (1997), says that the fauna of Curiaù is conducive to the formation of habitats and ecological niches, in the Cerrado and the forest islands there is an ornithofauna, passerines, rodents and reptiles. The flooded fields, streams and drainage channels develop a diverse ichthyofauna that forms the food base for the community's residents. Among the species we find: the Traira (Hoplias malabaricus), the Jéju (Hoplerythrinus unitaenia-tus), the Tambaqui (Colossoma macropomum), the Tamoatà

(Hoplosternum sp.), the Aracù (Leporinus sp.) and the Tucunaré (Cicha ocellaris).

According to Fagundes and Gibson (2000), Curiàu's agriculture is extensive subsistence and satisfies only the basic needs of their diet, using primitive and rudimentary techniques, limited to the cultivation of small areas. Residents practice slashing, burning and preparing the soil with hoes, and the authors believe that this misuse of the soil soon leads to its exhaustion. The main agricultural activity is the cultivation of manioc for the production of flour, on a small scale they grow vegetables, using mineral and organic fertilization techniques, irrigation and pesticides.

In Videira (2013) we find that livestock farming in Curiaù is more extensive in the form of buffalo farming, while cattle farming is currently practiced by small-scale farmers, as is pig and horse farming (see Figure 2) on a smaller scale. Fishing is the main extractive activity in Curiaù, followed by timber extraction and açai gathering. Hunting was once an abundant activity in the Curiaù community, practiced even by hunters who didn't live in the community, but with the current legislation this activity has been practically extinguished in Curiaù.

Figure 2: Horse breeding in Curiaù.

Source: Photographic collection of the authors.

Quilombola Communities in Brazil

It is estimated that between the years 1550 and 1850, four million black people arrived in Brazil, brought by force from the African continent, especially from the regions where they are located today: Guinea, Benin, Ivory Coast, Mali, Congo, Angola and Mozambique (ABRIL, 2013, p.310).

Blacks arrived in Brazil crammed into the holds of slave ships. Many died during the voyage, and those who survived were sold as slaves to work in agriculture and mining. For more than 300 years, slave labor was the country's main workforce and the basis of all economic activity.

In Ajayi (2011) we find that during the 19th century, until the end of the slave trade in 1870, out of a total of almost two million enslaved Africans, 1.1 million or 60% of them came to Brazil. In 1890, Brazil had an Afro population of around four million blacks, representing 33% of the local population and 36% of the continent's total Afro-American population. Olic and Canepa (2012) point out that the huge contingent of slaves brought to Brazil belonged to different cultures and religions on the African continent.

The formation of quilombos, which were communities formed by escaped slaves trying to survive outside colonial society in Brazil, was a movement of resistance to slavery. In the 20th century, when movements and organizations emerged to defend the rights of the black population, the Quilombo dos Palmares, founded by its leader Zumbi, emerged as a historical reference to these movements.

In Brazil, some black communities that originated from quilombos still survive, and today these communities are called quilombola communities. Since 1970, these quilombola communities have been identified in all Brazilian states by the Palmares Foundation, which is linked to the Ministry of Culture. Most of these communities are located in the North and Northeast (ABRIL, 2013, p.120).

The Brazilian Constitution of 1988 recognizes the right to land ownership of these black populations, who come from quilombos, and this process of recognition began in 1995. Most of these communities are made up of dozens of families, and some have thousands of inhabitants, mainly in the states of Maranhao and Bahia. Generally, these communities are located in isolated places. As they are quilombo remnants, they have a way of life in which collective land ownership, subsistence farming and animal husbandry predominate.

Since 2003, the criterion used to recognize a quilombola community has been self-identification, dispensing with the need to present documents proving descent from former slaves and uninterrupted possession of the territory. However, problems related to the demarcation of quilombola lands occur in several Brazilian states, where the residents of these communities often come into conflict with farmers and landowners in these places.

Figueiredo (2011) explains that the importance given to the quilombola issue is related to the broader emergence of the right to recognition of rural black communities. For the author, the perspective of multicultural recognition points to the guarantee of special rights, relating to their own language, religion, historicity and territoriality, but it can also imply differentiated access to social rights, such as education, work and security, requiring the state to apply affirmative policies.

The Quilombola Community of Curiaù

According to the State Coordination for the Articulation of Black Rural Quilombola Communities of Amapà (CONAQ/AP, 2014, n.p.), -there are 138 quilombola communities throughout the state of Amapà, located in all regions and around eight municipalities". There are currently more than fifty associations that make up CONAQ/AP, whose purpose is to fight for the emancipation of quilombola rights and to guarantee the preservation of land rights.

CONAQ/AP aims to work in defense of the rights of quilombolas in the state of Amapà, with the main goal of defending the quilombola territories of Amapà, the work developed in these 138 communities has the mission of strengthening their policies.

In Morais (2011) we find that the Curiaù community officially received the title of Quilombola Community on November 3, 1999, conferred by the Palmares Foundation, making it the first quilombola community recognized in the state of Amapâ.

-The African contribution has been present in Amapá society since the colonial period" (MORAIS, 2009, p.76).), when several Africans came to Amapâ, mixing and adapting to the existing cultural patterns, building and maintaining a culture that is still manifested today in religious festivals, music, cuisine, language and other artistic practices, such as the manifestations of: marabaixo, batuque, tambor, candomblé, capoeira, Iadainhas, processions, folias and traditions of the ancestors of the black community that are part of the cultural formation of Amapà.

According to Soares and Rodrigues (2008), it was through the organization of quilombos in places that were difficult to access that hundreds of blacks lived for many years as they had lived free in Africa. According to Morais (2009), as happened throughout Brazil, a group of black fugitives from the Belém region founded a quilombo on the banks of the Anauerapucu River in 1749. In the period between 1750 and 1782, the number of slaves brought to the region increased greatly. Blacks who did not accept slavery rebelled and fled, forming the quilombos of Maruanum, Igarapé do Lago, Ambé, Cunani, Curiaù and Goiabal.

Soares and Rodrigues (2008) tell us that the first African slaves arrived in Amapà in 1751, brought by families from the states of Rio de Janeiro, Pernambuco, Bahia and Maranhao, and then, with the introduction of rice cultivation, several slaves were imported from Portuguese Guinea (Africa). Portuguese Guinea was the name of present-day Guinea-Bissau, as a Portuguese colony between 1446 and 1974. Located on the west coast of Africa, Guinea-Bissau borders Senegal (to the north), Guinea (to the south and east) and the Atlantic Ocean (to the west). Also part of Guinea-Bissau's territory is the Bijagòs archipelago, made up of more than 80 islands.

Still in Soares and Rodrigues (2008) we find that during the construction of the Fortaleza de Sao José de Macapà, between 1764 and 1782, a large workforce was needed, and that Portugal had established several colonies in the north of the African continent in the 16th and 18th centuries, disputing this region with the Moors for centuries. Due to consecutive defeats, the Portuguese government transferred 340 families of Portuguese settlers who lived in the city of Mazagao (in the north of

Africa, now El Jadida), to Morocco. These families came with their slaves to Brazil to help defend the entrance to the Amazon River, settling in the territory of Amapa in 1770.

In Silva (2000), we find that the formation of the Curiaù quilombo began with the arrival of a couple of African origin and with them seven more slaves. According to Baena (1969), these first inhabitants came from Mazagao in 1771 to occupy the lands of Vila Sao José do Macapâ. While Marin (1997) informs us of the arrival of blacks from the city of Mazagao in 1760 in Macapâ.

According to Marin (1997), in 1760, the lands of Curiaù were handed over to these mazagonists, along with a hundred head of cattle to encourage cattle breeding, with the aim of taking advantage of the natural fields that existed on these lands.

Morais (2011) tells us that the origin of the name Curiaù is associated with one of the purposes of the area, which is cattle breeding (cria) and the mooing of cows (mu), resulting in the term *criamu*, which later became Criau and is currently called Curiaù. Their main crop is manioc, which is used to produce flour, and they also grow vegetables (lettuce, spring onions, coriander, cabbage, watermelon, passion fruit, lemons, oranges, avocados and others) on a small scale for local consumption. The marketing of açai is an activity carried out in the community, and açai and flour are also components of the local population's daily diet.

In Silva (2004), the author states that the people of the Curiaù community resisted and are resisting the strong influence of slavery in Amapà, especially in Macapâ. For the author, the steadfastness of these people in maintaining their sovereignty, culture, customs, habits, experiences and beliefs comes from their persistence in believing in educational and professional growth, with a view to preserving the environment, in order to guarantee the survival and future of the community's residents. The struggle to guarantee their heritage means that cultivation is a major source of income for the Quilombola community of Curiaù.

Located 8 kilometers from Macapâ, the capital of Amapà, the community of Curiâu maintains customs and traditions stemming from the descendants of slaves who were refugees some two centuries ago. The site is an Environmental Protection Area (APA) and was granted the title of quilombola community in 1999 by the Palmares Foundation for preserving the traditions maintained since the arrival of the African couple and seven other refugee slaves. In 2007, the mayor of the city of Macapâ, Joao Henrique Pimentel, accepted the request of the residents of Curiaù and through a letter sent to the Electoral Court requested the suspension of a plebiscite to transform Curiaù into a district, as established in the municipal legislation of 2004, which provided for the administrative redivision of Macapâ. But the residents of Curiaù didn't accept this, because for them Curiaù already has many achievements, the most important of which was the recognition of the quilombo area, a cultural heritage, which for them would be de-characterized by the change.

At a public hearing held on October 21, 2007 in Curiaù, the residents of the community protested against the change and asked for the plebiscite not to be held, through a document containing 600 signatures that was delivered to Macapà's Planning Secretary Alfredo Ramalho, who passed on the request to the mayor, who decided to respect the will of the residents, Alfredo Ramalho, who passed on the request to the mayor, who decided to respect the will of the Curiaù residents and suspend the plebiscite, after which a bill was drafted and sent to the City Council asking for the change to Macapà's master plan.

In Amapà (1992), we find that the Curiaù APA was created by State Decree No. 1417, of September 28, 1992, with the aim of protecting and conserving the area's natural and environmental resources.

According to Visentini, Ribeiro and Pereira (2013),

> [...] in traditional African society, the notion of village chief strictly meant representation, that is, the chief was a delegate of the people. In this sense, traditional African society didn't place the individual above the people, but rather linked them to the group, making everyone stand together in a complex structure of interdependence. (VISENTINI; RIBEIRO; PEREIRA, 2013, p.27).

One of the residents of Curiaù, Mr. Eraldo, told us that the leader of this community is Mr. Joaquim. Mr. Eraldo also told us about the community and stressed the importance of preserving and valuing the heritage left to the younger residents, and that another resident, Mr. Sebastiao, was concerned about this and had written seven books telling the story of the community, three of which we had access to (SILVA, n.d., 2000, 2004).

One of the quilombola community's greatest challenges is to keep traditions alive in the region. To this end, part of the population is mobilizing to pass on the roots of the Curiaù people to younger people, who today have more contact with other cultures. This can be seen during dance performances, which include ladies and children, but few teenagers. During the interview, Mr. Eraldo recalled what he had experienced in Vila de Curiaù and made a point of passing on some of his experiences to the younger generation, encouraging all forms of culture in the region:

> [...] Here, anyone from outside who wants to build their own house needs to marry a brunette from here, or a brunette. Then they'll have the right to make their own territory and be in contact with our culture and adapt to it. (ERALDO).

Mr. Eraldo also told us that the 1,500 residents of the community are related by blood.

Culture and Religion in the Curiaù Community

According to Silva (2004), the life story of the people of Curiaù has, over the years, resisted and preserved the most precious thing that the Curiaù community possesses: its culture. A remnant of the former Afro-Brazilian Quilombo, the Curiaù community, made up predominantly of Afro-descendants, has preserved its customs and cultural traditions, arousing interest and attracting people to attend its celebrations of the saints. The culture of Curiaù is linked to religion, because ever since the first inhabitants of the community arrived in Curiaù from Mazagao, they have treated religiosity as something natural. Festivities in the community are seen as an obligation, where litanies are prayed and offered to the saints as a way of giving thanks. They celebrate Saint Sebastian on January 19 and 20, Saint Lazarus on February 10, 11 and 12, Saint Joseph on March 19, Saint Mary on May 30 and 31, Saint Anthony from June 12 to 14, Saint Joachim from August 9 to 19, Saint Raymond from September 6 to 8, Our Lady of the Conception from December 7 to 9 and Saint Thomas from December 20 to 22, preserving the integrity of their ethnocultural values and roots.

For Videira (2013), faith in the saints and religious festivals are of great importance in the lives of the people of Curiaù, and the author believes that these festivals are part of the affirmation of the community's identity. The festivals of saints of the Catholic tradition held in the community of Curiaù are called Batuque and Marabaixo. For the residents of Curiàu, taking part in these festivities is a source of pride and joy, where they share moments of reunion with their families, relatives, friends and neighbors, where they celebrate their ancestors, rescuing during these festivities, through orality recorded and incorporated into the memory of the elders of Curiaù, the local tradition that is taught by the older residents to the younger ones, who learn to listen, tell and remember the "causos" of the past and relate those of the present, making the absence and presence felt in the memory of those who have already left, continuing the tradition of the Afro-Amapa people in reverence for their devotional saints and preserving the history of their ancestors.

Saint Joachim, the Patron Saint of the Curiaù Community

Silva (2004) states that the Catholic religion has always been part of the life of the residents of Curiaù. The author tells us that the most important religious festival held in the community is that of Sao Joaquim, the patron saint of Curiaù. In Videira (2013) we find that of all the festivals on Curiaù's Afro-religious and cultural calendar, the festival in honor of Saint Joachim is the one that receives the most attention, being called "Batuque do Glorioso Sao Joaquim". The author says that the residents of the community feel proud and responsible for maintaining the tradition of celebrating this Batuque in honor of Sao Joaquim, where the litanies, prayers of revelry, dawn, dawn, payment of promises, celebration of masses, parades, games, dances, preparation and tasting of food and drink are carried out.

According to Videira (2013), during the days of celebration, food and drink are distributed free of charge to people: meat (beef and pork, chickens and ducks), cozidao (stew with vegetables and beef), cake, tapioquinha, corn porridge, fruit (pineapple, açai, banana, Brazil nut, apple, mango, watermelon, melon, orange, pear, tangerine, grape and uxi), fruit juices (acerola, bacuri, cupuaçu, graviola, orange, mango, passion fruit, muruci and taperebà), tea, chocolate, soft drinks, gingibirra (the traditional aphrodisiac drink of the festival, made from ginger, cloves, cachaça, water and sugar) and wine. In Silva (2004) we find that, during the São Joaquim festivities, children and teenagers are baptized, masses are celebrated in the morning and processions in the afternoon.

Traditional and Cultural Festivals of Curiaù

Silva (2004) explains that Batuque and Marabaixo have a very strong meaning for the inhabitants of Curiaù. Batuque is the biggest manifestation of the community, it is sung by several local people, the special moment of Batuque is Aurora, which is sung at five o'clock in the morning. Those who are still asleep wake up to mingle with those who are singing to strengthen the celebration, everyone feels proud, with a sense of accomplishment for having taken part in the Batuque.

Still in Silva (2004), we find that Marabaixo begins with the cutting down or taking down of a pole, made from a tree called "pau-espirito-santo", whose bark is medicinal, found in the community itself. This pole is placed in the place where the Marabaixo will be held, which can be a house, a headquarters or a shack. During the afternoon of the festival, a crowd goes into the forest to break the branches of the muteiras (trees considered sacred to the community), the rameiras (the branches of the muteiras) decorate the pole, wrapped around it, and on top of the pole is placed a flag with the image of Santa Maria, Mother of Jesus or other saints that are being celebrated. It's a very joyful festival, with people arriving on horseback, bicycle and on foot. The participants sing the "ladrào" (Marabaixo or Batuque song).

Videira (2009) explains that the songs of Marabaixo are composed of verses that are called -ladrào", these verses have the objective of criticizing, exalting, thanking, lamenting or satirizing facts that occur in the daily life of the community and in social relations. According to the author, in the lyrics of the Marabaixo songs we can see the presence of Afro-descendant literature, the old songs are still sung and faithfully maintained, just as the old creators of the verses composed them. For the author, the Marabaixo songs are historical documents that represent a form of expression and resistance, a critique of the black people of Amapá in the face of the impositions, racist and discriminatory attitudes practiced against the Afro-Amapá people, against their culture and against their history.

For Videira (2009, p.139) -Afro-descendant cultural expressions" such as Marabaixo from the north and northeast of Brazil must be (re)known and valued as legitimate representatives of Afro-Brazilian cultural diversity, the author also presents the lyrics of some Marabaixo songs, so that we can understand their meanings, within the historical and social records expressed in their verses and rhymes of the thieves.

Below are excerpts from two songs that are widely sung in the community:

Refrain - *Cafusa my Cafusa/What have you come to do here/I've come to get my things/I don't care about you*

Ladrao - *I came to get my things/ I don't care about you/ You thought I wasn't coming/ Goodbye, I'm leaving* Refrao - *Cafusa minha Cafusa*

Ladrao - *I wish I could see you dead/ And the vultures eating you/ And your bones on the tray/ On the street being sold*

Chorus - My Chorus...

Ladrao - *I have a pain in my chest/ And another in my heart/ How much I miss/ My great passion*

Refrao - *Cafusa minha Cafusa...* (VIDEIRA, 2009, p. 140-141).

Ladrao - *Menina se queres vamos/No se pôr a imaginar/Quem imagina cria medo/Quem tem medo no vai lá.* (VIDEIRA, 2009, p. 142).

Videira (2009) explains that the songs of Marabaixo express the memories of the Afromapans in the verses of the thieves, showing the subjectivity in the verses, bringing the historical subject closer to the social space remembered. For the author, the events that took place in the community were recorded in the memories of the older residents, recounting the daily life (re)constructed by the community's residents. For this author, the Marabaixo songs represent the memory of the ancestors, being passed on from generation to generation through orality, making it possible to record the history of the community's residents.

For Maciel (2001), the preservation of Marabaixo is an important black cultural manifestation that demonstrates the ethnic identity of the black people of Amapá.

ETNOMATEMATICS

Monteiro (2004), when talking about Mathematics Education, discusses the need to relate everyday knowledge to school knowledge, and highlights Ethnomathematics as a philosophical proposal that debates the production, validation and legitimization of mathematical knowledge in different social practices. The know-how of a group can be understood as a "cultural product", created by this group in its daily work and transformed by the interaction that emerges from contact with other groups.

Still in Monteiro (2004), we find that the knowledge present in everyday practices, as well as mathematical knowledge, is constituted within a group. This knowledge, interpreted and created by the group itself, is different from school knowledge. For this reason, the author warns us of the need to create spaces for this knowledge to also be present in the school context, enabling students and teachers to critically appropriate the different forms of knowledge of those involved in the teaching and learning processes. For the author, the school must incorporate the idea that the legitimization of knowledge can be found in the autonomy and will of a group that legitimizes this knowledge on the grounds of its coherence and applicability, ensuring that teaching and learning processes are connected to different fields of knowledge, providing knowledge of differences, without belittling or denying what is their own.

The aforementioned author believes that Ethnomathematics seeks to situate knowledge in its historical and cultural context, also valuing values not legitimized by the school, allowing the legitimization of knowledge from other groups and social practices, articulations between everyday and school knowledge, since the school has the social commitment to reproduce the values legitimized by society at a given historical moment.

Federal Law 10.639/03 and Ethnomathematics

Education is a broad and complex process of social, cultural and technical knowledge and actions, oriented towards knowing how to be and how to live together in multiracial societies such as Brazil, seeking to educate and train citizens for equality and respect for differences in the field of attitudes and knowledge. Pedagogical actions to combat racism are aimed at respect for differences and ethnic-racial awareness, combining practices in the field of attitudes and knowledge to implement Federal Law 10.639/03 with the aim of bringing social interactions between individuals and groups into Brazilian education.

The principles of Law 10.639/03 guide an education that is free of racism and promotes equal rights, guided by an understanding of the anthropological, cultural and geographical diversity that surrounds the different ethnic-racial groups, creating educational bases that value and respect the diverse material and immaterial contributions of these groups.

Trigo (2013) says that Law 10.639/03 is the result of the struggle of groups that defend the rights of black and Afro-descendant people and is justified as an attempt to correct historical and conceptual distortions, often repeated in school discourse, and that for these groups these distortions have contributed to ongoing social injustices. For the author, it is imperative to debate education at the service of diversity, with the challenge of affirming, revitalizing and valuing the self-image of black people.

Brazil (2004), when referring to educational actions to combat racism and discrimination, says that education systems and Basic Education establishments, at the levels of Early Childhood Education, Elementary Education, Secondary Education, Youth and Adult Education and Higher Education, need to include in the content of subjects and in curricular activities "knowledge of African origin and/or that concerns the black population" (BRASIL, 2004, p. 24). For example: in Mathematics, contributions from African roots, identified and described by Ethnomathematics.

Trindade (2008), when referring to the recognition of the importance of ethnic-racial issues, questions where the African heritage that makes up Brazil is being made visible and puts us before the challenge of (re)understanding the black presence in Brazil and in the African diaspora. The author states that there is a certain lack of knowledge about the African heritage that marks our Brazilianness, since we are unable to observe this heritage/influence in social sectors, the arts, science and technology.

Brasil (1997), when referring to Ethnomathematics, says that among the works that have gained expression in the last decade, the Ethnomathematics Program stands out, with its alternative proposals for pedagogical action. This program opposes orientations that disregard any closer relationship between mathematics and socio-cultural and political aspects - which keeps it untouched by factors other than its own internal dynamics. From an educational point of view, it seeks to understand thought processes, ways of explaining, understanding and acting in reality, within the individual's own cultural context. Ethnomathematics seeks to start from reality and arrive at pedagogical action in a natural way, through a cognitive approach with a strong cultural foundation.

Mathematics Education and Ethnomathematics

According to Mendes (2009b), Mathematics Education, as an area of study and research, is made up of a body of activities aimed at developing, testing and disseminating innovative teaching methods; designing and implementing curricular changes; creating and testing support materials for the teaching and learning of Mathematics.

Mathematics Education is also fundamental in the continuing education of mathematics teachers, with the aim of making teaching more effective and fruitful, in order to overcome the difficulties encountered by teachers and students during the educational process at the different levels of basic education.

Also according to Mendes (2009b), mathematical thinking is a human construction developed within a historical-social context with reflections and applications from this same context, and needs to be widely understood by all mathematics educators. We know that many have dedicated themselves in recent decades to the development of studies that have served as a basis for the construction of a theoretical framework on which to base educational actions.

Within this context, a large number of mathematical educators, reflecting on the philosophical assumptions and pedagogical practices of Mathematics Education, have managed to emerge methodological guidelines for the realization of a more meaningful Mathematics Education, thus giving rise to methodological trends in Mathematics Education with their characteristics, their pedagogical principles and their modes of approach. Pointing out various possibilities for using each of them, according to the needs of the teaching and learning processes.

The Trends in Mathematics Education arose from the need to find solutions to some of the obstacles encountered by mathematics educators in the course of their practice. The possibility of helping to improve the pedagogical practice of teachers based on teaching experiences made these trends a reality: the use of concrete materials and games, Ethnomathematics, Problem Solving, Mathematical Modeling, the History of Mathematics, the use of computers and calculators in mathematics teaching and the Didactics of Mathematics.

Also in Mendes (2009b) we find that Ethnomathematics has a socio-cultural and cognitive approach and is one of the fields of Mathematics Education that has aroused a lot of interest among scholars, researchers and educators who are looking for solutions to problems related to the epistemology of mathematics and its teaching. Being conceptualized as the confluence zone between mathematics and cultural anthropology, it can be considered an area of knowledge linked to cultural groups and their interests. Recognizing that all cultures and peoples develop ways of explaining, knowing and dealing with their realities, in the search for this understanding there is a need to: quantify, compare, classify and measure, which makes mathematics emerge spontaneously.

Ethnomathematics, Afroethnomathematics and Cultural Plurality

Next, we will relate the concepts derived from Ethnomathematics to enable discussion of the historical contributions of the black population to the construction of Brazilian society and our African heritage, how these contributions can be discussed during mathematics classes, so that students can have a more appropriate dimension of the contribution of black people to the construction of the country, addressing the issue of cultural plurality.

D'Ambrosio (2011) encourages us to reflect more broadly on mathematical thinking, to try to understand mathematical knowledge/doing throughout human history, not by proposing a new epistemology, but rather -to understand the adventure of the human species in the search for knowledge and the adoption of behaviors" (D'AMBROSIO, 2011, p. 17). This author highlights the need to always be open to new approaches, new methodologies, new visions of what science is and how it has evolved. Science itself develops the intellectual tools for its critique and for incorporating elements from other knowledge systems. These intellectual tools depend heavily on a historical interpretation of the knowledge that lies at the origins of modern knowledge.

D'Ambrosio (2011) states that while the subordination of disciplines and scientific knowledge itself distances education from its objective of prioritizing the human being and their dignity as a cultural entity, Ethnomathematics has a very natural relationship with Anthropology and the Cognitive Sciences, and its anthropological dimension is evident, as it is linked to communities, societies and cultural groups that identify with common goals and traditions. Furthermore, the political dimension of Ethnomathematics is indisputable, as it is focused on recovering the cultural dignity of human beings, which is violated by social exclusion and the discriminatory barriers established by society.

D'Ambrosio (2011) emphasizes that we should see Ethnomathematics as a new field of research, a proposal that presents itself as a research program on the history and philosophy of mathematics, with important repercussions on education. The author understands -mathematics as a strategy developed by the human species throughout its history to explain, to understand, to manage and to live with sensible, perceptible reality, and with its imagination, naturally within a natural and cultural context". And he says that he sees Education as a -strategy to stimulate individual and collective development" generated by cultural groups, with the aim of maintaining themselves as groups and advancing in satisfying the needs of survival and transcendence, for him, consequently, -Mathematics and Education are contextualized and interdependent strategies" (D'AMBROSIO, 2011, p. 83).

In D'Ambrosio (2011, editor's note) we find that the movement called Mathematical Education is based on -the principle that everyone can produce Mathematics in its different expressions". Ethnomathematics, considered a trend in mathematics education, seeks to deepen and "analyze the role of mathematics in Western culture and the notion that mathematics is only one form of ethnomathematics".

We realize that Ethnomathematics goes far beyond the discussion of race and ethnicity, as it is close to Afroethnomathematics, discussed by Cunha (2004), who identifies Afroethnomathematics as the area whose main concern is the cultural resources that facilitate the learning and teaching of mathematics in areas with a majority Afro-descendant population. According to this author, Afroethnomathematics studies the contributions of Africans and Afro-descendants to mathematics and computer science, as well as developing knowledge about the teaching and learning of mathematics, physics and computer science in Afro-descendant majority territories.

For Santos (2011), the discussion of Afroethnomathematics should not be left out of any discussion on the relationship between African culture and Mathematics Education for the Brazilian population, as this exclusion may imply a less in-depth discussion of the scope of this topic, stating that Afroethnomathematics studies African history and the mathematical evidence found in the various African cultures; in addition to working with evidence of mathematical knowledge in African religious knowledge, popular myths, constructions, arts, dances, games, astronomy and mathematics itself, carried out on the African continent, extending to the areas of African diaspora.

In Brazil, this field of study emerged from the development of pedagogical practices by the Black Movement, in the quest to improve the teaching and learning of mathematics in quilombo remnant communities and in urban areas where the majority of the population is of African descent, known as the black population.

The aforementioned author states that the concern with the teaching and learning of mathematics in areas with a majority Afro-descendant population is due to the precariousness of formal education. The subject of mathematics in these places, where quality teaching is practically non-existent, has a shortage of teachers, a precarious structure and content, and the students don't see themselves represented in the classes of this curricular component.

For Santos (2011), the most serious aspect of this situation is that students' failure in this subject is not attributed to the education system, but rather to themselves, with ideas about their inability to learn mathematics subtly appearing between the lines.

In Silva (1999, p. 15-16) we find that "curriculum is place, space, territory. Curriculum is a relationship of power. The curriculum is a trajectory, a journey. [...] Curriculum is text, discourse, document. The curriculum is an identity document".

> [...] the knowledge that constitutes the curriculum is inextricably, centrally and vitally involved in what we are, in what we become, in our identity, in our subjectivity. Perhaps we could say that as well as being a question of knowledge, the curriculum is also a question of identity (SILVA, 1999, p. 15-16).

Rethinking the integrated curriculum with a view to deepening the theoretical-methodological approach to education for ethnic-racial relations means understanding the construction of human subjectivity and making it possible to include Afro-Brazilian and African culture, it means valuing the history and culture of Afro-Brazilians and Africans, it means contributing to the implementation of Law 10.639/03.

Costa and Oliveira (2010) argue that the teaching of mathematics should be geared towards a better understanding of reality, social phenomena and the development of citizenship, contributing to socio-historical transformations. Therefore, the subject of Mathematics allows us to reflect on cultural and racial diversity, and Ethnomathematics can contribute significantly to the dissemination and social appreciation of African and Afro-Brazilian history and culture, if we consider the implementation of Law 10.639/03 as an important measure that can, in addition to changing a situation of institutional racism, lead students to realize the cultural, social and political dimensions of Mathematics.

We understand that Ethnomathematics, being a research program in the history and philosophy of mathematics, involves ethnic-racial issues in its breadth, and can promote discussion/reflection on the potential for implementing Law 10.639/03.

According to Oliveira (2011, p.1)

> The theme of *Cultural Plurality* seeks to value and respect the ethnic and cultural characteristics of different social groups, providing students with the possibility of making a broad reading of Brazilian diversity [...]. In the context of Mathematics Education, there is the possibility of thinking about interdisciplinary work, which appropriates this search, through the Ethnomathematics Program.

Brazil (1997), when dealing with Cultural Plurality, states that the construction and use of mathematical knowledge is not only done by mathematicians, scientists or engineers, but, in different ways, by all socio-cultural groups, who develop and use skills to count, locate, measure, draw, represent, play and explain, depending on their needs and interests. Valuing this mathematical, intuitive and cultural knowledge, bringing school knowledge closer to the cultural universe in which the student is inserted, is of fundamental importance for teaching and learning processes. On the other hand, by giving importance to this knowledge, the school contributes to overcoming the prejudice that Mathematics is knowledge produced exclusively by certain social groups or more developed societies. In this work, the History of Mathematics, as well as Ethnomathematics studies, are important for explaining the dynamics of the production of this knowledge, historically and socially.

Frankenstein and Powell (1997) and Knijnik (1996) state that Ethnomathematics recognizes that all cultures have produced and produce mathematical knowledge, and considers it relevant to include this knowledge in the school curriculum so that it can be contemplated and understood in its diversity, in accordance with the vision of Cultural Plurality, pointed out by the National Curriculum Parameters - PCN.

Brasil (1997) suggests that each school develops projects involving issues related to ethnic-racial relations, racial diversity and cultural plurality that are considered relevant to the community. Themes related to education and cultural diversity, for example, are privileged contexts for developing content that establishes a historical-cultural relationship with the numerical sense, records of the primitive process of counting, measurement, percentages and the monetary system, legitimizing the African origins of knowledge, highlighting Afro-Brazilian civilizational values. As well as the idea of symmetry, which is related to that of harmony and proportion, according to Brasil (1997), the objectives of work with this theme are to identify symmetry in plane figures, sensitivity to observing symmetry in nature, in the arts, in buildings and the transformation of a figure in the plane by means of reflections, translations and rotations, which may deserve special attention in Mathematics planning, as they are present in African and Afro-Brazilian cultural manifestations.

Claudia Zaslavsky's work is noteworthy. Her book, published in 1973, was a pioneer in recognizing that many of the mathematical practices found in Africa have their own characteristics; it is a true ethnomathematics, although the term was not used (D'AMBROSIO, 2011, p. 24).

In this sense, for Oliveira (2011), ethnomathematics can enhance and energize the implementation of Law 10.639/03, because for the author:

The law should not be seen as a new discipline or methodology to be employed, but as a possibility for new dialogues and new attitudes, in order to provide the emergence of a transformative education, in relation to ethnic-racial discrimination, in all disciplines of the school curriculum. (OLIVEIRA, 2011, p.3)

Gerdes (2010) informs us that in different cultural environments, on all continents, women and men decorate objects, create shapes and patterns, in an artistic-mathematical way. Teachers can work on this type of knowledge and share it with their students during math classes, valuing the different cultural mathematical knowledges, relating the subject of mathematics to the cultural universe of African and Afro-Brazilian origin, contributing to a mathematics education without any ethnic-racial discrimination.

Interest in the ethnomathematics of African cultures has grown enormously. The work of Paulus Gerdes and his collaborators in

Mozambique, with a large number of publications in Portuguese, mainly analyzing basketry, textiles and traditional games in southern Africa (D'AMBROSIO, 2011, p. 24).

In order to establish new dialogues and ethnic-racial relations, the math teacher can use the concepts of Ethnomathematics and relate mathematics to other subjects such as history, arts, philosophy, religion, literature, Portuguese language, physical education and biology, enhancing the immersion of African and Afro-Brazilian culture in the school space, working on ethnic-cultural diversity. In this way, the teacher would be taking into account the historical and socio-cultural processes worked on in these or other subjects, because, according to Costa (2009), the ideas, knowledge and actions related to classification, inference, ordering, explanation, modeling, counting, measurement and spatial and temporal location originate, live and are renewed from the needs that a group of people feel in relation to their survival and transcendence, This always takes place in a historical and cultural context that is inseparable from the language used by the group, the codes of behavior adopted, social practices, values, myths, rites, knowledge modified or learned through the cultural dynamics of the encounter, the power relations established between the group and nature, between the people of the group itself and between the group and other groups, the art and religiosity of the group itself, as well as other collectively shared knowledge and cultural manifestations.

> The recognition of mathematical practices in everyday life in Africa has been the subject of important research. A very interesting example is the use of percussion instruments, an integral part of African traditions. The rhythm that accompanies percussion instruments can be studied as an aid to understanding reason (D'AMBROSIO, 2011, p. 24).

In this sense, we understand that Ethnomathematics points the way to the relationships that can be established between Afro-Brazilian civilizational values, playfulness, memory, ancestry, circularity and orality, proposed by Trindade (2006), who presents some didactic-pedagogical proposals in Mathematics that can be worked on in line with the guiding principles of the National Curriculum Parameters - PCN, with regard to valuing ethnic-cultural diversity, with the intention of giving students the opportunity to learn about, recognize and highlight Afro-Brazilian civilizational values, linking mathematics, culture, education and society.

For Trigo (2013), -despite the recommendations that African and Afro-Brazilian themes permeate all subjects" (TRIGO, 2013, p.38), the guidelines of Law 10.639/03 themselves recognize that -discussions can be extremely productive in History, Literature and Arts classes" (TRIGO, 2013, p.38). The author emphasizes that it is necessary to -address the African contribution to the formation of Brazilian identity, beyond what is established" (TRIGO, 2013, p.38) in traditional curricula. For her:

> [...] it is a question of removing the black person from the position of victim and situating them as a historical subject, capable of resisting and imposing the force of their culture" (TRIGO, 2013, p.38). In literature and the arts, it is essential not only to show how Africans and Afro-descendants have been represented by official culture, but also to highlight their importance as cultural producers, in music, in circles more restricted to official culture - predominantly white and Eurocentric in character - there has been a black presence (TRIGO, 2013, p. 38).

-Despite the apparent difficulty in addressing ethnic and racial issues" (TRIGO, 2013, p.38) in science and mathematics, the teaching of these subjects -can help to deconstruct the idea that Africans contributed little to the development of scientific knowledge. Africa was the cradle of humanity, and consequently of human knowledge" (TRIGO, 2013, p.38). The author also states that:

> African peoples created tools, techniques and work systems that contributed, for example, to the development of agricultural production and mineral exploitation. This knowledge has spread to various regions of the world. In Brazil, for example, enslaved blacks introduced the use of the hoe in agriculture and the beater in gold mining. In addition, it has recently been discovered that some African peoples, including the Egyptians, mastered iron metallurgy three centuries before the Europeans (TRIGO, 2013, p. 38-39).

Still in Trigo (2013), we find that in order to implement Law 10.639/03 in the school curriculum, for this author, the first thing to do is to start guiding students, showing them where the introduction of black culture in Brazil began, and where our history begins. For the author, "the change in the approach to Afro-Brazilian themes will require the involvement and commitment of the entire school community. The school as a whole will have to come up with strategies so that the racial issue is no longer just another aspect of the curriculum" (TRIGO, 2013, p.39), but in fact becomes real knowledge that transforms society.

Africa the Anthropological Cradle of Humankind

We know that Africa is the anthropological cradle of humanity. Eves (2004) tells us that the first peoples inhabited the open spaces of the savannahs of Africa, southern Europe and Asia, and Central America. These peoples were nomads, moving around in search of food and according to climatic changes, they inhabited the Earth during the period known as the *Stone Age*, between 5,000,000 BC and 3,000,000 BC, when Australopithecus, an ancestor of man who lived in Africa, produced stone axes and knives. Later, in 4,000,000 BC, Homo erectus sought shelter in caves to protect himself from the temperatures of the savannahs.

According to Eves (2004), from 110,000 B.C. to around 35,000 B.C. "Homo neanderthalensis heated his caves with fire, cooked the animals he caught in the savannahs" (EVES, 2004, p. 22) and recorded his hunts in paintings on the cave walls. Around 30,000 BC, Homo sapiens migrated from caves to mobile tents, which he made from animal skins and covered with wood, which he could take on hunting trips. This new man carved statues and religious icons in stone.

Historically, we usually place the end of the Stone Age at 3,000,000 BC, but when the conquistadors arrived in southern Africa between the 16th and 17th centuries, most of the peoples they encountered were still living in the Stone Age. And some cultures persisted in the Stone Age in some places until the 19th or 20th century.

According to Eves (2004), in the Stone Age period the scientific and cultural advances recorded were limited due to the fact that all people were nomadic hunters. This period has been divided into three periods by historians: Paleolithic (or Old Stone Age, from 5,000,000 BC to 10,000 BC), Mesolithic (or Middle Stone Age, from 10,000 BC to 7,000 BC) and Neolithic (or New Stone Age, from 7,000 BC to 3,000 BC). After these three periods, the Stone Age began to decline, ushering in the Bronze and Iron Ages, when hunter-gatherer peoples began to develop agriculture, fruit picking, domestication and animal husbandry.

Historians consider the Paleolithic period to be a period of transition from a society formed by pre-humans to a society of human hunters, while the Neolithic period is also considered to be a period of transition from a society of hunters to a society of farmers. Stone Age society did not have the equipment to smelt metals, so they did not develop metal tools, cities, science or written language.

Eves (2004) states that in 20,000 BC these savannah hunters developed a complex culture creating tools, language, religion, art, music and commerce, but progress in science and mathematics encountered obstacles in the social and economic structure of that period. Because the peoples were hunters and not farmers, they moved according to the seasons and carried tools, clothing and personal artifacts.

The lack of scientific progress in the Stone Age led to barter activities, where it was necessary to keep records of these hunts and exchanges, and these activities required the idea of counting. Some peoples had pictographic calendars where they recorded their stories for decades. These primitive counting systems only evolved with the emergence of agriculture in the last millennium of the Stone Age, in the Neolithic period, because agricultural activity needed arithmetic, and society in this period went from collecting fruit, roots and vegetables, to planting and cultivating seeds, and harvesting crops.

The primitive peoples who were hunters and gatherers became farmers, they did not develop scientific traditions, but after 3,000 BC the communities formed by farmers, who populated the banks of the River Nile in Africa, created cultures that began to develop science and mathematics.

According to Eves (2004), Mathematics began in the Stone Age, with pre-humans who showed a numerical sense and recognized very limited models of numerical and spatial relationships with animals, birds, insects, plants, in the crystallization of minerals, in the phenomena of nature, in the trajectories of planets and comets, considering the efforts of primitive man to systematize concepts of quantities, shapes and numbers as the oldest manifestations of Mathematics, with the concept of number and the process of counting emerging.

Eves (2004) states that there are archaeological records that man in 50,000 BC was able to count, and that this counting process occurred in a natural and conjectural way, appearing even before the first historical records, primitive man had some numerical sense, recognizing when animals, fruit, food, tools and objects produced in the Stone Age were put in or taken out (more or less).

According to Eves (2004), as society evolved during this period, simple counting processes became unavoidable, as it became necessary for man to know if the quantity of animals, food or objects he possessed was decreasing. This hypothetical development is supported by reports from anthropologists who have studied primitive peoples, such as the "Ishango bone", over 8,000 years old, found in Ishango, Zaire, on the shores of Lake Edward, which shows numbers preserved by carvings on it.

Boyer (1996) says that mathematics originally emerged as part of man's everyday life, and for him the persistence of the human race is probably related to the development of mathematical concepts. He believes that at first the primitive notions of number, magnitude and form could be related to contrasts rather than similarities, and from this perception of these dissimilarities in number and form, science and mathematics were born. The author explains this perception better when he states that:

This perception of an abstract property that certain groups have in common and which we call number, represents a major step on the road to modern mathematics. It is unlikely that it was discovered by an individual or a tribe; it is more likely that it was a gradual perception, developed as early in man's cultural development as the use of fire, perhaps 300,000 years ago (BOYER, 1996, p. 1).

Still in Boyer (1996), we find that:

For the prehistoric period there are no documents, so it is impossible to follow the evolution of mathematics from a specific drawing to a familiar theorem. But ideas are like resistant seeds, and sometimes the presumed origin of a concept can just be the reappearance of a much older idea that has been forgotten. Prehistoric man's preoccupation with configurations and relationships may have originated in his aesthetic sense and the pleasure he took in the beauty of shapes, motives that often drive today's mathematics. [...] We can make conjectures about what led Stone Age men to count, measure and draw. That the beginnings of mathematics are older than the oldest civilizations is clear. To go further and categorically identify a specific origin in space and time, however, is to confuse conjecture with history. It is better to suspend judgment on this question and go further, on the firmer ground of the history of mathematics found in written documents that have come down to us. (BOYER, 1996, p. 5).

For Boyer (1996) -the thousands of years it took for man to distinguish between abstract concepts and repeated concrete situations show the difficulties that must have been experienced in order to establish a basis, albeit a very primitive one, for mathematics" (BOYER, 1996, p. 4). The author believes that statements about the origin of mathematics are necessarily risky, since the beginnings of the subject are older than the art of writing. For information about prehistory, we depend on interpretations based on the few artifacts that remain, evidence provided by anthropology and retroactive, conjectural extrapolation from the documents and records found.

Giordani (2013) says that archaeological finds have, due to their nature, something objective about them.

a special significance as indicators and measures of civilization: iron objects and the technology involved in their manufacture, ceramics with their production techniques and styles, glass pieces, writings and graphic styles, navigation techniques, fishing and weaving, food products and also geomorphological, hydraulic and plant structures linked to the evolution of the climate (GIORDANI, 2013, p.41).

Dating processes contribute to broadening and deepening our knowledge of African history.

Mazrui and Wondji (2011, p.761) show that, -because of colonialism, the contribution of African scientists to human knowledge has been modest since 1935", but the authors state that it is not appropriate to measure science only by the activity of scientists, because -history is not made by historians, but by society". In the author's view, "in the same way, scientific elaboration is not due solely to scientists, but to the whole of the community". We can "highlight the reasons why African society became one of the pillars of Western science and technology" (MAZRUI; WONDJI, 2011, p.761-762), due to colonization, "although colonialism hindered the development of science and technology in Africa" (MAZRUI; WONDJI, 2011, p.762). This same colonial condition was "a transmission link for African material contribution to science and technology on a Western scale" (MAZRUI; WONDJI, 2011, p.762).

For Mazrui and Wondji (2011) there is a force in Africa greater than the colonial experience, and that force is African culture. For these authors, the study of trends in African science and technology recognizes the prominence of values and traditions relating to African philosophy and science, "considering knowledge as an empirical phenomenon" (MAZRUI; WONDJI, 2011, p.762). Using Marxist terms, African science and philosophy are part of the superstructure, "with culture playing the role of the base or infrastructure", because "science and philosophy transcend geographical space and historical time", they are not limited to the geographical boundaries of Africa and neither are they limited to "the historical period that began in 1935". In order to understand "African science and philosophy, their strengths and their limits", we must reposition them in their social and cultural context (MAZRUI; WONDJI, 2011, p.763).

According to Aranha and Martins (2009), for Marx, society is structured on two levels: the first level, known as the infrastructure, is the economic base, encompassing man's relations "with nature in an effort to produce his own existence and those of the environment".

(ARANHA; MARTINS, 2009, p. 323); the second level, called the superstructure, is of a political-ideological nature. In Aranha and Martins (2009) we find that the Marxism elaborated by Karl Marx (1818-1883) and Friedrich Engel (1820-1895) emerged in Germany in the 19th century.

Marx and Engels formulated their theories based on the social reality they observed.

For Marx, it is a mistake to analyze human beings in isolation from their reality, which consists of all social relations. According to the authors cited above, for Marx, matter is a primary factor, the source of consciousness is a secondary factor, because consciousness derives from matter, that is, consciousness is a reflection of matter.

For Aranha and Martins (2009) we need to differentiate Marxist materialism, which is dialectical, from mechanistic materialism, because mechanistic materialism is based on the observation that the world is made up of material particles that combine inertly, whereas in dialectical materialism material phenomena are processes. For the authors, Marx inverts the process of common sense, which explains history by the actions of individuals, by applying the principles of dialectical materialism to the field of history, from Marx's historical analysis, thus giving rise to historical materialism.

In order to study society, according to Marx, one should not start from what individuals say, imagine or think, but from the way in which they produce the goods necessary for their existence. By analyzing the relationships that these individuals establish with nature in order to transform it through work and the relationships between them, we discover how they produce their lives and their ideas.

According to Aranha and Martins (2009, p. 325), -Marx calls human action transforming reality praxis. This concept is not properly identified with practice, but signifies the dialectical union of theory and practice", while -consciousness is determined by the way existence is produced, human action is also projected, reflected, conscious and capable of modifying theory". For the authors, a

> [...] the fundamental category for understanding dialectics is totality, whereby the whole predominates over its constituent parts. This means that things are in constant reciprocal relationship, and no phenomenon of nature or thought can be understood in isolation from the phenomena that surround it. Facts are not atoms, but belong to a dialectical whole and as such are part of a structure. [...] Dialectics is the contradictory structure of reality, which in its constitutive movement passes through three phases: thesis, antithesis and synthesis. In other words, the movement of reality is explained by the antagonism between the moment of thesis and antithesis, whose contradiction must be overcome by synthesis. (ARANHA; MARTINS, 2009, p.323).

Still in Aranha and Martins (2009), we find that - in analyzing the social being, Marx develops a new anthropology, according to which there is no identical "human nature" in every time and place". The individual is formed as he modifies nature through work, and since work is based on collective action, then "the human condition depends on his social existence" (ARANHA; MARTINS, 2009, p. 324, emphasis added). Marx explains this process through the concepts of production relations, productive forces and mode of production.

For Marx, the relations of production reveal the way in which human beings use techniques to organize themselves through the division of social labour, starting from their natural conditions. The forces produced are the "ensemble formed by the climate, water, soil, raw materials, machines, labor and instruments of work." When the instruments of work change, the productive force undergoes alterations that cause changes in the forms of relationship between individuals. And the mode of production is "the way in which the productive forces are organized in certain relations of production at a given historical moment" (ARANHA; MARTINS, 2009, p. 324).

For Aranha and Martins (2009, p. 325), one of the consequences of changes in the productive forces is changes in production relations and the mode of production, for example, "the slave mode of production stems from the increase in production beyond what is necessary for subsistence", requiring the use of new labor forces, transforming human beings into slaves, "an example of the first form of human exploitation".

Santos (2013) argues that in order to understand what happened, we must take into account the complexity of the racial issue. For this author, Brazil was a country that lived through the experience of slavery for more than three hundred years, in which the participation of Africans and their descendants in the constitution of Brazilian society was based on this initial condition of slavery. For her, the process of abolishing slavery did little to change the condition of social marginality of former slaves and their descendants.

França and Ferreira (2012) explain that the struggle in the Quilombo dos Palmares for freedom from oppression took on a Marxist coloration and the Quilombo dos Palmares emerged as an example of what moves humans on earth according to Karl Marx's disciples: the class struggle. For Aranha and Martins (2009), the result of the contradiction established by the slave regime is expressed in the class struggle. The class struggle takes place when the productive forces develop to a certain point and reach an advanced stage. These productive forces "dialectically enter into contradiction with the old relations of production, because they become inadequate" (ARANHA; MARTINS, 2009, p. 325), and so the need arises for a new division of labor.

The various Africas

Pennaforte (2009) says that, in order to talk about Africa, we must be aware of the divisions to which the African continent is subjected, paying attention to the criteria that guide these regionalizations so that we can understand the reality of this continent.

The first major division separates Africa into two: North Africa (or Mediterranean Africa) and Sub-Saharan Africa, with the Sahara Desert as the natural divider of the continent, which is divided as follows: Mediterranean Africa is made up of the following countries: Egypt (African), Libya, Morocco, Tunisia, Algeria and Western Sahara.

Sub-Saharan Africa is made up of: Guinea-Bissau (formerly Portuguese Guinea), Guinea, Sierra Leone, Ivory Coast, Liberia, Gambia, Senegal, Mauritania, Mali, Cape Verde, Burkina Faso, Ghana, Togo, Benin, Nigeria, Niger, Chad, Sudan, Central African Republic, Cameroon, Congo, Gabon, Equatorial Guinea, Sao Tome and Principe, Angola, Democratic Republic of Congo, Uganda, Kenya, Ethiopia, Djibouti, Eritrea, Rwanda Burandi, Kenya, Somalia, Tanzania, Comoros, Madagascar, Mauritius, Malawi, Zambia, Zimbabwe, Mozambique, Botswana, Namibia, Swaziland, Lesotho and South Africa.

Pennaforte (2009) explains that:

> To the north of the desert are the countries bathed by the Mediterranean Sea (which is why the region is called Mediterranean Africa). Forming an "arc" from the Atlantic to the Red Sea, Mediterranean Africa includes Western Sahara, Morocco, Algeria, Tunisia, Libya and Egypt. The various countries to the south of the Sahara make up the sub-Saharan portion (PENNAFORTE, 2009, p.9).

Still in Pennaforte (2009) we find that it is common to associate Mediterranean Africa with -White Africa" and Sub-Saharan Africa with -Black Africa", expressions used in the 19th century, which for the author are impregnated with racist conceptions, which can give us the mistaken impression that the whole of these societies is homogeneous, which is a mistaken impression.

Pennaforte (2009) reports that if we consider the natural, economic and social characteristics of the African continent, Africa can be divided into five regions: the Northern Region, the Western Region, the Central Region, the Southern Region and the Horn of Africa Region (see Figure 3).

Figure 3 : Division of Africa into 5 Regions.

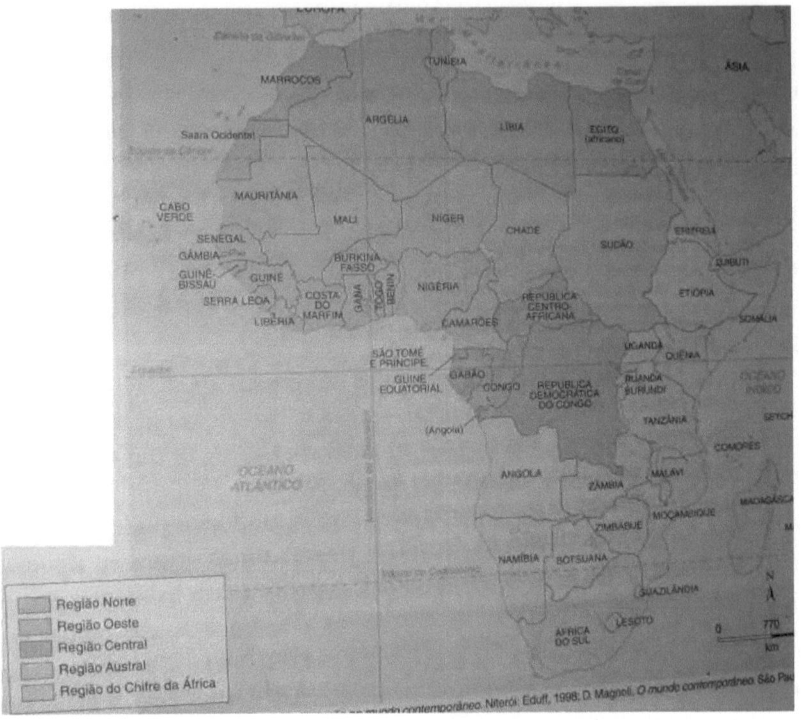

Source: Penaforte (2009, p.9).

The Northern Region is made up of: Western Sahara, Morocco, Algeria, Tunisia, Libya and Egypt (African).

The Western Region is made up of: Cape Verde, Senegal, Gambia, Guinea-Bissau (formerly Portuguese Guinea), Guinea. Sierra Leone, Liberia, Ivory Coast, Ghana, Togo, Benin, Nigeria, Cameroon, Burkina Faso, Mauritania, Mali, Niger, Chad and Sudan.

The Central Region consists of: Central African Republic, Gabon, Congo and Democratic Republic of Congo.

The Southern Region consists of: Comoros, Mauritius, Madagascar, South Africa, Lesotho, Swaziland, Namibia, Angola, Botswana, Zimbabwe, Zambia, Mozambique, Malawi, Tanzania,

Rwanda Burandi, Uganda and Kenya.

And the Horn of Africa Region by: Eritrea, Djibouti, Ethiopia and Somalia.

For Pennaforte (2009), Southern Africa is the region with the best economic conditions. In the Northern Region, the most striking aspect is the concentration of population along the Mediterranean coast. The Western Region is rich in minerals, with diamond, iron, oil, phosphate and copper mines, but is also marked by civil conflicts. In the Central Region there are forests similar to the Amazon, which hinders the development of agriculture, this region is also marked by civil wars due to mineral resources, diamond deposits. And finally, the Horn of Africa, where the desert climate prevails, is a region plagued by famine, drought and war. In this region, the United Nations Organization (UNO) is constantly called upon to alleviate the suffering of the population.

THE SCHOOL IN THE QUILOMBOLA COMMUNITY OF CURIAÙ

-It takes a whole village to raise a child" (African proverb)

The José Bonifàcio State School is located in the Quilombola community of Curiaù. The school was founded in 1943, but moved to the building where it currently stands in 1992.

This public school was officially founded in 2001, but its history goes back to the 1940s. In Amapà (2001) we find that Decree No. 0197 of January 23, 2001, published in the Official Gazette of the State of Amapà on January 24, 2001, made official the creation and naming of the José Bonifàcio State School in the municipality of Macapà.

The school offers primary education from the 1st to the 9th grade, in the morning from 7:30 am to 12:00 pm and in the afternoon from 1:30 pm to 6:00 pm. It also has a class of 3ª Stage of Youth and Adult Education - EJA from 13:30 to 18:00.

The school has one (see Figures 4 and 5):

- 01 Board;
- 01 Secretary;
- 01 Teachers' room;
- 03 bathrooms (01 male bathroom, 01 female bathroom and 01 bathroom for teachers and staff);
- 01 sports court;
- 01 kitchen;
- 01 Covered area (where school meals are served);
- 01 patio;
- 09 classrooms;
- 01 tutoring room;
- 01 Teaching room;
- 01 School TV room;
- 01 reading room;
- 01 recreation and games room;

- 01 Computer laboratory;
- 01 Library.

Figure 4: School façade, corridors and library.

Source: Photographic collection of the authors.

Figure 5: Area where lunch is served, kitchen and Pedagogical Support room.

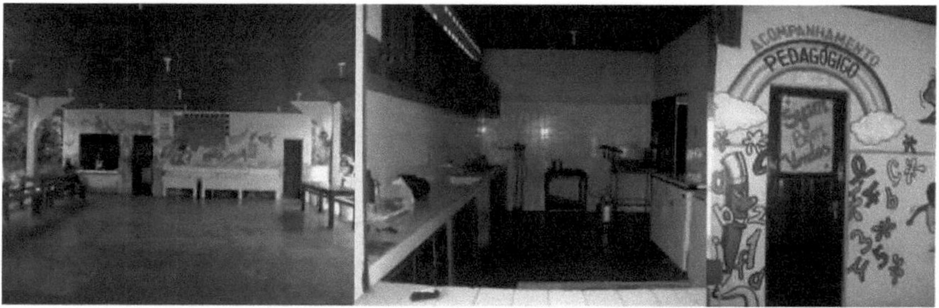

Source: Photographic collection of the authors.

In 2014, the school served a total of 275 students who lived in or around the community. When the students finish elementary school, they are enrolled in schools in the city center. The school receives funding from the state government, which helps the students by providing river and land transportation (boat and bus) free of charge for all the students enrolled at the school, according to the list sent by the school principal to the State Department of Education.

The teachers and school staff who live in the community itself are called "children of the community" and the other teachers and staff live in neighborhoods very close to Curiaù such as: Jardins, Jardim Felicidade, Açai, Infraero I and II, Brasil Novo, Boné Azul, Pedrinhas, Goiabal and Renascer.

Videira (2013) states that education is a citizen's right, and that quilombola education is also a right, but it needs experiences that are suited to the specific needs of the community, in terms of the way the residents live and their cultural practices, which makes quilombola education a challenge.

For the author, the school in the Curiaù community has an innovative pedagogical project, created by her, by a group of teachers and school managers who met, analyzed, experimented, dared and reflected on the appropriate educational approach for this community's school, which incorporates the history of Curiaù and brings it into the classroom, as a form of an educational process designed and carried out for the students who are part of this community. For this author, the work carried out in this school has been significant and significant in the lives of the students and residents.

During our research, we observed that the teachers contextualize their lessons with the day-to-day activities of the community's residents. These residents, who are parents of students at the school, contribute to school meals by donating food grown on their own properties. These farming parents donate fruit, vegetables, spices and small animals to help prepare the daily meals their children eat at school.

The content of the lessons (see Figure 6) is based on Law 10.639/03, which, as we said earlier, amends Law 9.394, of December 20, 1996, which establishes the guidelines and foundations of national education, determining that in the official curriculum of the school network it will be compulsory to include syllabus content referring to the study of the theme "Afro-Brazilian History and Culture", the struggle of black people in Brazil, black Brazilian culture and black people in the formation of national society, rescuing the contribution of black people in the social, economic and political areas, pertinent to the History of Brazil, which will be taught within the scope of the entire school curriculum.

Figure 6: Students' productions in class.

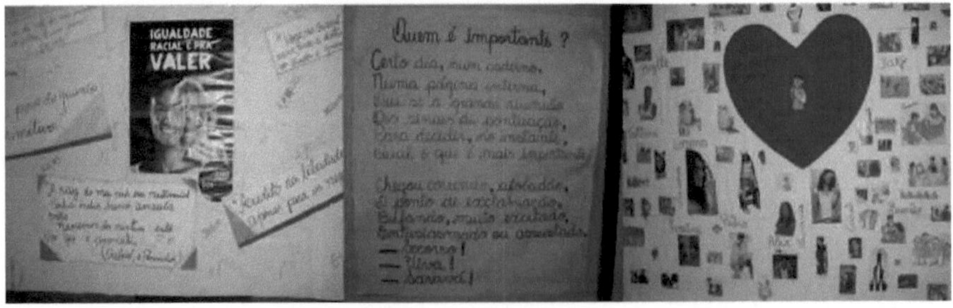

Source: Photographic collection of the authors.

The school premises also contain posters and murals depicting African and Afro-Brazilian culture (figures 7, 8 and 9).

Figure 7: Posters displayed in the school corridors.

Source: Photographic collection of the authors.

Figure 8: Library, classroom mural and sports court.

Source: Photographic collection of the authors.

Figure 9: Photo of the mural on the Library wall.

Source: Authors' photographic collection

The school runs various projects that seek to involve students and encourage them to stay at school longer. Among these projects, we would like to highlight the main ones, which are: the *More Education Program*, *School Reinforcement*, *Oral Knowledge*, *Music and Percussion*, *French classes*, the *Arts Project*, *Hair Braiding* and the *Curiaù Shows Your Face Project*. All these projects have as their central axis the rescue of the culture of Afro-descendant values, the valorization of black people and the integration of African culture into the daily life of the community and the school.

There are other interesting and award-winning projects at the school. One of these is developed by teacher *Vanda*, called *Mala Màgica*. Using theater, drawings, puppets, storytelling, literature and other resources, the teacher interacts with the children and deals with important educational values, always highlighting and valuing Afro-Brazilian culture. This project began three years ago when the teacher would visit the classes with a suitcase full of surprises and, based on what she took out of the suitcase, she would build a story with the students during the teaching and learning processes, an interesting construction of knowledge in a playful way. There is a lot of receptivity and the teacher now receives the classes in a room full of teaching resources, everything in the suitcase is used at a given time.

The Curiàu school is also home to the country's first quilombola orchestra, which is part of RankBrasil. Made up of 45 students, the musical group performed for the first time on October 25, 2012, at the José Bonifàcio State School itself, with the participation of the Fire Brigade band, as well as performances of ladainha (a poetic form) and marabaixo (a dance typical of the state).

The school orchestra is made up of 45 students, but a total of 100 people are involved in this project, which provides musical initiation with wind, wood, string and classical percussion instruments, without losing the regional culture. The orchestra's proposal is to teach music as a tool for social inclusion, giving people who live in the quilombola community and also in neighboring communities access to education and culture, as well as the possibility of entering the job market.

The José Bonifàcio State School seeks to value Afro-descendant culture, with projects that emphasize batuque, another manifestation of local customs. The students know how to dance carimbó, drum and sing the songs of the saints. Relying on the willingness of the teachers, the school's management decided to include French classes in the curriculum from the 1st to the 9th year of elementary school. -French is because of our border with French Guiana. This will help the children a lot in the job market," explained teacher Claudeci Ferreira da Silva Rodrigues.

The *Curiâu Mostra Tua Cara (Curiâu Shows* Your *Face*) Project is based on the guidelines of Law 10.639/03. It has been carried out since 2003 by art teacher Irene Bonfim, and is a project to value the roots, history and culture of the Curiaù community,

According to Videira (2013), the Project's guidelines are:

- Evidence of the practical implementation of Law 10.639/03;
- Turning some of the content to the reality lived and experienced by Criauenses and students in the school environment;

- And to give visibility to the history and culture of the Curiaù quilombo.

At the end of 2009, the *Curiaù Mostra Tua Cara* Project won a prize in a competition promoted by the Ministry of Culture in the area of Popular Cultures. The José Bonifàcio State School received a prize worth ten thousand reais to be invested in the project's activities.

Math classes at Curiaù School

Brazil (1997), when referring to educational actions to combat racism and discrimination, says that education systems and Basic Education establishments, at the levels of Early Childhood Education, Elementary Education, Secondary Education, Youth and Adult Education and Higher Education, need to include in the content of subjects and in curricular activities "knowledge of African origin and/or that concerns the black population" (BRASIL, 1997, p. 24). For example: in Mathematics, contributions from African roots, identified and described by Ethnomathematics.

According to Costa (2009), ideas, knowledge and actions related to classification, inference, ordering, explanation, modeling, counting, measurement and spatial and temporal location originate, live on and are renewed based on the needs that a group of people feel in relation to their survival and transcendence. This always takes place in a historical and cultural context that is inseparable from the language used by the group, the codes of behavior adopted, social practices, values, myths, rites, knowledge modified or learned through the cultural dynamics of the encounter, the power relations that are established between the group and nature, between the people of the group itself, between the group and other groups, from the art and religiosity of the group itself, as well as other collectively shared knowledge and cultural manifestations.

Trindade (2006) presents us with some didactic-pedagogical proposals in Mathematics that can be worked on in line with the guiding principles of the PCN, with regard to valuing ethnic-cultural diversity, with the intention of giving students the opportunity to get to know, recognize and highlight Afro-Brazilian civilizational values, linking mathematics, culture, education and society.

In accordance with the recommendations of Brasil (1997), at the Curiaù school, when planning math classes, the teachers relate the content to African and Afro-Brazilian cultural manifestations (see Figures 10, 11 and 12). For example, in 5th grade classes, when working on math content in the classroom, the teachers use information about African countries, such as population, territorial extension, demographic density, flags, etc. This data is researched in books and atlases available in the school library.

Figure 10: 5th grade class C and 5th grade class B.

Source: Photographic collection of the authors.

Figure 11: Fifth grade class A.

Source: Photographic collection of the authors.

Figure 12: Math teacher and 9th grade students and library collection.

Source: Photographic collection of the authors.

ACTIVITY HELD AT THE CURIAÙ SCHOOL

During the course of this research, a workshop was held at the Curiaù community school using the African game Mancala. The workshop was attended by the 4th grade class, made up of 17 students, 9 girls and 8 boys, and teacher Dalva.

Holding this workshop at the José Bonifàcio State School contributes even more to the daily contact with the residents of the Curiaù community (Afro-descendants who are remnants of the former Afro-Brazilian Quilombo), staff, teachers and students of this educational institution, providing the opportunity to observe and participate in a pedagogical practice in the classroom, in another school context.

During the workshop, we were able to experience the daily lives of the community's residents and workers, the school's staff, students and teachers, learn more about the teaching methodology used by the teachers and their relationship with the students.

This workshop took place in two stages: the first was observation, during which we watched classes from the 4th, 5th and 9th grades and learned more about the school's administrative and pedagogical workings. The second practical moment was when we held a workshop using the Mancala game. Teacher Dalva took an active part in the workshop and at the end of it, she made a critical analysis of the game, as she was aware of the research being carried out.

Our aim was to carry out a mathematics activity at the school, in accordance with the guidelines of Law 10.639/03. To experience a more direct contact with the teachers, students and school staff and to observe and describe what would happen during the workshop with the students.

Profª . Dalva explained to us that at the start of the school year the class had 15 students, because the school principal really tries to do a different job with classes of only 15 students, but some parents go to the Public Prosecutor's Office asking for places for their children to study. According to the teacher, when the Public Prosecutor's Office calls the Department of Education and the latter checks the system for available places, the school that always appears to have places available is the José Bonifàcio State School. This is why three of the school's classes, which started the school year with 15 pupils, had 17, 23 and 29 pupils.

The Mancala Game

In Mendes (2009a), we find that the use of concrete materials and games in the teaching of mathematics is a didactic alternative that helps the teacher to intervene in the classroom during their activities, individually or in groups, making the student an active agent in the construction of their own knowledge.

The guidelines proposed in the National Curriculum Parameters for Secondary Education - PCNEM, in relation to games say that they are very valuable elements in the process of appropriating knowledge, and that "they should be used in the disciplines of Nature Sciences, Mathematics and their Technologies, because they offer stimulation and a favorable environment that encourages the spontaneous and creative development of students" (BRASIL, 1999, p.141).), allow teachers to broaden their knowledge of active teaching techniques, "stimulate students' ability to relate to school content in a playful, enjoyable and participatory way, leading to greater appropriation of the knowledge involved" (BRASIL, 1999, p.141).

Souza (2003) says that a game in general has a challenge to overcome, rules to follow, players who give it life and uncertainty about the outcome, which is the reason for continuing to play. Brougère (1998, p. 122) states that when games are used in class with the aim of providing -acting, learning, educating without knowing through recreational exercises", they are called educational games. This type of game can provide the construction, development or retention of content, because every game has a problem or a challenge, and the main objective is to get the player to find ways to solve it, by means of different possible moves and eliminating those that hinder the achievement of the result.

According to Muniz (1999), the analysis of moves helps to understand the error and makes it possible to get it right. Each move gives rise to an exchange of opinions, or rather, in each move, cognitive capacities are mobilized and the player displays both their acquired knowledge and their ability to create new strategies. It can be said, then, that the game provides intellectually active work, as well as socialization, solidarity and affection.

According to Santos (2008), the African game Mancala dates back some 7,000 years and its origins lie on the African continent, more precisely in Egypt. Its oldest boards were found in excavations in the Syrian city of Aleppo, in the Karnak temple (Egypt) and in Theseum (Athens). From the Nile valley, it spread throughout Africa and the East. Today it is played on all continents and spread by its fans and educators in schools and universities.

Zaslavsky (2000) tells us that -Mancala" is an Arabic word meaning -transfer". This game was spread worldwide by enslaved black Africans and there are more than 300 ways to play it, and the rules of the game vary according to the region where they are played. According to Ripoll [s/d] the word Mancala is also used to indicate games that have the following characteristics: played in pairs, each player has a series of holes in a line where the seeds are distributed, and it is not said to move the seeds but to sow them.

According to Rêgo (2000), -it enables action planning, sequencing, manipulation of quantities, exploratory action, develops logical reasoning and also makes it possible to work with the operations of addition and subtraction" (RÊGO, 2000, p.150). It can be used from early childhood education to higher education.

In Lima, Gneka and Lemos (2005) we find that Mancala is a strategy game related to sowing. It comes from the Arabic word *nagaala,* which means "to move". It simulates the act of sowing, the germination of seeds in the soil, their development and harvest. According to the authors, the movement of the seeds across the board was associated with the celestial movement of the stars, and the board itself symbolized the Sacred Arch.

In its early days, Mancala had a magical meaning, related to sacred rites. In some places, the games were reserved only for men or priests. Today, in most countries, Mancala has lost its magical and religious character. However, the Alladians in Côte d'Ivoire retain the religious meaning and believe that Mancala can only be played in sunlight. At night, they offer the boards to the gods to play.

Proof of the importance of this game for the Alladians is the need for a game of Mancala between the contenders for the throne, so that the king's successor can be chosen. Another interesting fact is that the game of bùzios, associated with candomblé, is derived from Mancala.

The Activity

Mancala trays can be made from alternative materials such as clay, wood, MDF, cardboard, E.V.A., egg cartons, cardboard or styrofoam. The holes in the tray can be made with a stylus or scissors.

Eighteen kits were made for the workshop (see Figure 13), containing:

- 1 tray;
- 1 Envelope with 48 bean seeds;
- 1 Mini-manual containing the *Rules of the Game.*

Figure 13: Contents of the kits made for the workshop.

Source: Photographic collection of the authors.

The costs of purchasing the material to make the 18 kits, with 2014 figures, are listed in the table below:

Table 1 - Expenditure on material for making the kits

Material	Unit Cost (R$)	Quantity	Total Cost (R$)
E.V.A. sheet	1,50	12	18,00
Tube of E.V.A. Glue	1,30	2	2,60
Envelope	0,10	18	1,80
Transparent packaging	0,40	18	7,20
Stylus	2,10	1	2,10
Scissors	3,00	1	3,00
Lapis	0,40	1	0.40
Beans (Kg)	6,50	½ Kg	3,25
Copy of the rules of the game	0,07	9	0,63
TOTAL (R$)	-	-	**38,98**

Source: Prepared by the authors, based on the acquisition of the material to make the kits.

As previously reported, the workshop with the Mancala game was held with the students and the teacher of the 4th grade class. On the day of the activity, the 18 kits were distributed, one for each student and one for the class teacher (see Figures 14 and 15).

Figure 14: Conducting the workshop

Source: Photographic collection of the authors.

Figure 15: Workshop.

Source: Photographic collection of the authors.

At the start of the workshop, we explained to the students that there are various ways of playing Mancala, but that the rules used would be those of Awalé, which is one of the denominations of Mancala. The students were informed about the existence of other denominations of Mancala in their respective countries, listed below:

1. Adi - Dahomey;
2. Andot - Sudan, especially by the Bega tribe and is thrown on the ground;
3. Aware, Awalé, Awari - Upper Volta, Suriname, Gulf of Guinea;
4. Ayo - Nigeria;
5. Baulé - Ivory Coast, Philippines and Sound Islands;
6. Kakuà - Ghana, Nigeria;
7. Kalah - Algeria;
8. Oware - Ghana, played especially by the famous Ashanti;
9. Tantam - Apachi;
10. Walu, Adji and Ti - Brazil;
11. Wari - Sudan, Gambia, Senegal, Haiti;

We told the students that according to Lima, Gneka and Lemos (2005), Mancala or Awalé is the fruit of the ideas, the way of thinking and also the collective memory of the African people. For these authors, through Awalé the player gets to know the African soul or that of the Baobab, because in some African countries it is with their grains that they play. The game has one foot in mythology and the other in everyday life in Africa, because when we play, what we are doing is repeating the cycles of nature: the cultivation of the soil and the harvests, which follow the rhythm of the seasons. Originating in the north of the Gulf of Guinea, Awalé began to travel around the continent, then the world, until it arrived in Brazil.

The students also learned that every game has a strategy, and that the rules can vary slightly according to the region of Africa where it is played. Mancala or Awalé is based on the continuous redistribution of seeds, sowing to reap is the fundamental principle, which does not vary. This is the secret and the source, in ancestral African practice, of exchange.

The strategies used in Mancala are mathematical calculation exercises through which players can develop their mental speed, logic and concentration. All in a playful way. According to Lima, Gneka and Lemos (2005, p. 56) "Awalé or Mancala is an easy game to learn, and years of playing can make certain players invincible, but above all it is a game based on generosity, because to win a player has to know how to give to his opponent".

The students were told that the game is always played by two players who share a single board, and that some of them should keep their games so that they can play with one of their classmates (see Figures 16 and 17).

Figure 16: The workshop.

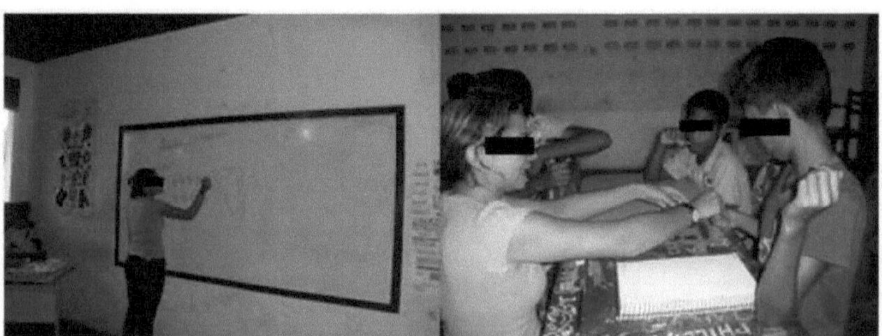

Source: Photographic collection of the authors.

Figure 17: Workshop.

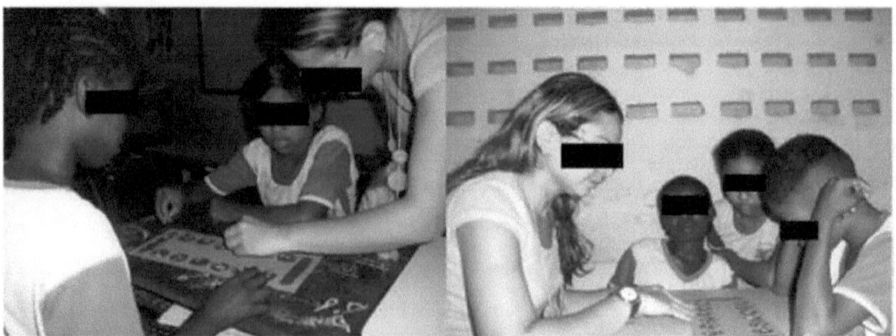

Source: Photographic collection of the authors.

We gave the students an oral reading of the rules of the game (which each student received in the kit), written by Georges Gneka and quoted below:

1 **- Aim of the game: to** harvest a large crop. The player who harvests the most seeds by the end of the game wins.

2 **- The course** is divided into 2 territories, each with 6 holes. The holes are called pits. Each player chooses their territory: south or north. Each player chooses their territory, south or north, and stands facing the holes in their territory.

3 **- Start of the game:** each hole will receive 4 seeds equally, so that each player fills all the holes in their field, planting 24 seeds in total.

4 **- Each player's turn:** the participants agree who will start the game. Whoever starts chooses one of the pits in their territory and takes from their set of seeds (4) to redistribute them.

5 **- The redistribution:** the direction of the game is always to the right. After emptying the chosen hole, the player places each of the 4 seeds in each of the following holes. Therefore, the 4 holes to the right of the empty one will each receive a seed.

6 **- Planting in the opponent's territory:** the next player to play is the opponent. In the same way, he chooses a hole in his territory, takes all four seeds from it and redistributes them, respecting the direction (always to his right) and the sequence (don't skip any holes). In this way, the seeds move between the holes in your territory, but also in those of your opponent. Each hole accumulates new seeds which are added to the initial seeds. Sharing also leads to situations in which the pits may run low on seeds.

7 **- Harvesting:** pits with 1, 2 or 3 seeds are at risk. If a player calculates well, so that the last seed distributed falls into an opponent's pit with 1, 2 or 3 seeds, he has the right to empty the pit, collecting the seeds for himself and taking them out of the game. But this only applies to pits with 3 seeds or less.

8 **- Multiple harvesting:** the opponent's pits with few seeds become targets. When one of the players manages to "harvest" all the seeds from a pit, as described above, all the preceding pits that also contain 1 to 3 seeds can be emptied. The player can thus manage to collect a large number of seeds from several pits in a row in a single turn.

9 **- Making a krou: making** a complete turn. If the hole chosen by the player to start the move has more than 11 seeds, he will have to place the rest in sequence, one in each hole, which will make him go all the way around the board, passing through both fields. In this case, with each pass the

player must skip the starting box, which must always be empty.

10 **- Feed:** in this game, you don't have the right to starve your opponent. If the opponent has no more seeds in his field, the other player must give him a seed, taken from one of his holes, so that the game can continue. From one seed you can get back many seeds.

11 **- Game over:** the game ends when the number of seeds is so small that no player can capture the other's seed. The player with the most seeds wins.

12 **- The player**: He plants and harvests seeds. He must calculate, based on the number of seeds he starts with, where they will fall and how much he will be able to harvest from his opponent. In the same way, he must calculate so that the holes in his territory are not short of seeds.

It was explained to the students that the game ends when there are so few seeds left on the board that no more can be captured. The player who has captured the most seeds will win the game.

At the end of the workshop, the students were asked to answer 5 questions about the workshop. These questions and the answers of two students are transcribed below:

Q: Did you enjoy the activity?

A: Yes, because it's cool.

Q: What did you think of the Mancala game?

A: Nice, because we had to eat each other's seeds.

Q: Did you understand the rules of the game?

A: Yes, you had to plant and eat the seed.

Q: Did you use math during the game?

A: Yes, I did the math.

Q: What did you like best?

A: I liked playing. (Pupil A - 11 years old.)

Q: Did you enjoy the activity?

A: Yes, because it's cool.

Q: What did you think of the Mancala game?

A: Cool, because we plant seeds.

Q: Did you understand the rules of the game?

A: Yes, because you have to eat the other person's seed.

Q: Did you use math during the game?

A: Yes, sums.

Q: What did you like best?

A: To play. (Pupil B - 09 years old.)

As mentioned above, teacher Dalva took an active part in the workshop and at the end of it, she wrote an analysis of the game in which she wrote the following: "The game presented is a great game, very important for the students, because it arouses their curiosity, interest, dedication and love of mathematics, teaching them to share, to help others. It was very rewarding. Thank you!" (DALVA).

Positive and Negative Aspects Observed

Brasil (1997) suggests that each school develops projects involving issues related to ethnic-racial relations, racial diversity and cultural plurality that are considered relevant to the community. Themes related to education and cultural diversity, for example, are privileged contexts for developing content that establishes a historical-cultural relationship with number sense, records of the primitive process of counting, measurement, percentages, the monetary system, legitimizing the African origins of knowledge, highlighting Afro-Brazilian civilizational values. During the workshop, we observed that the school in the quilombola community does this all the time, but that the teachers and staff involved in the projects don't have the habit of making photographic or audiovisual records of the events that take place in the school, nor do they make written reports of these events.

One of the difficulties encountered during the research and during the workshop was precisely the lack of this important information, which is not in the school archives, because we always asked about it, and the information we got was that the only person who has these records is the teacher Rosàlia, coordinator of the project - Curiaù Mostra a Tua Cara".

At the school office, we were told that we could find more information in the master's thesis and doctoral dissertation of a professor at the Federal University of Amapà - UNIFA, Professor Piedade Lino Videira, whose works we have cited in this study.

After the workshop, we believe that we have achieved the objectives set out in the workshop objectives, as the students talked a lot among themselves (see Figures 18 and 19) about how to capture seeds, and came to the conclusion that the last square where the player sowed must satisfy two conditions:

First condition: that it belongs to the opposing team;

Second condition: that it contains 1, 2 or 3 seeds, not counting the one just sown.

In this case, the player takes the seeds from that square, and the seeds from the previous square, as long as it also meets the conditions. And the seeds of the second preceding square, and so on, until they reach a square that no longer meets the conditions, at which point the game ends and the captured seeds remain with the player who captured them.

Figure 18: The workshop.

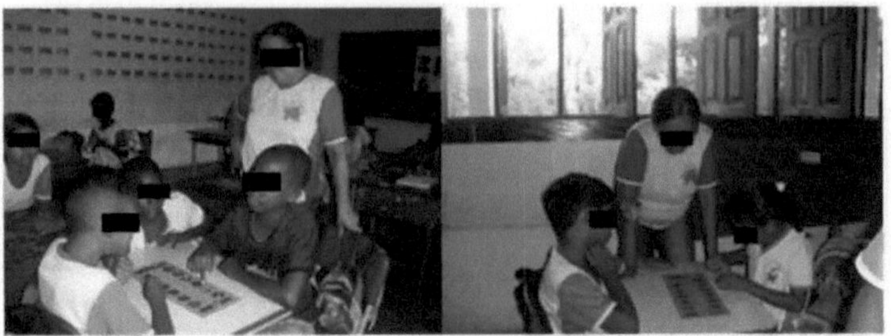

Source: Photographic collection of the authors.

Figure 19: The workshop.

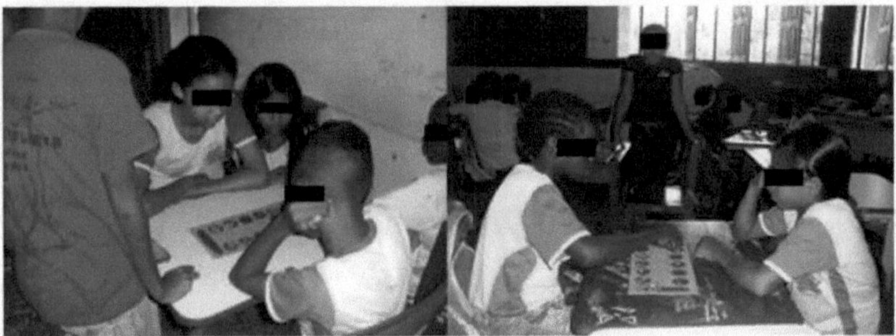

Source: Photographic collection of the authors.

Finally, everyone has understood the most important rule of the game: the player cannot leave the opponent's field without seeds. If they do, they must deposit a seed on their opponent's unseeded field.

As mentioned above, at the end of the workshop the students were asked to answer 5 questions about the workshop, but of the 17 students who took part in the workshop, only 9 gave us this activity, 5 of whom were female and 4 male.

THE FLOUR PRODUCTION PROCESS

Brasil (1997), when addressing Cultural Plurality, states that the construction and use of mathematical knowledge is not only done by mathematicians, scientists or engineers, but, in different ways, by all socio-cultural groups, who develop and use skills to count, locate, measure, draw, represent, play and explain, depending on their needs and interests. Valuing this intuitive and cultural mathematical knowledge, bringing school knowledge closer to the cultural universe in which the student is inserted, is of fundamental importance for the teaching and learning process. According to D'Ambrosio:

> Everyday life is impregnated with cultural knowledge and practices. At all times, individuals are comparing, classifying, quantifying, measuring, explaining, generalizing, inferring and, in some way, evaluating, using the material and intellectual tools that are specific to culture (D'AMBROSIO, 2011, p. 22).

On the other hand, by giving importance to this knowledge, the school contributes to overcoming the prejudice that Mathematics is knowledge produced exclusively by certain social groups or more developed societies. In this work, the History of Mathematics, as well as Ethnomathematics studies, are important for explaining the dynamics of the production of this knowledge, historically and socially.

Brasil (1997) suggests that each school develops projects involving issues related to ethnic-racial relations, racial diversity and cultural plurality that are considered relevant to the community. Themes related to education and cultural diversity, for example, are privileged contexts for the development of content that establishes a historical-cultural relationship with the numerical sense, records of the primitive process of counting, measurement, percentage, monetary system, legitimizing the African origins of knowledge, highlighting Afro-Brazilian civilizational values.

Frankenstein and Powell (1997) and Knijnik (1996) state that Ethnomathematics recognizes that all cultures have produced and do produce mathematical knowledge, and considers it relevant to include this knowledge in the school curriculum so that it can be contemplated and understood in its diversity, in accordance with the vision of Cultural Plurality, pointed out by the PCN.

According to Mattos and Brito:

> Fieldwork is full of mathematical knowledge, giving us the opportunity to cross the boundaries of the classroom in order to get to know our students' reality and thus understand the difficulties they face at school when applying content that is far removed from their context. (MATTOS; BRITO, 2012, pp. 969-970).

The 9th grade math teacher at the Curiaù school uses figures and numbers from the community itself when developing the content, such as the amount of flour produced and sold at the market and at the fairs in the city center, the costs of this production and the profit made on the sale. This process extends to the production and sale of tucupi, the harvesting and sale of açai, and other fruits such as acerola, pineapple, orange, lemon, mango, watermelon, maracuja, muruci and tapereba. Among the vegetables, the residents of the Curiaù community, who are farmers, plant and sell lettuce, cabbage, spring onions, green smell, okra, and manioc itself, the root from which they extract tucupi and farinha d'àgua. The costs of planting and the profit made from selling these vegetables are also used as examples during math lessons.

During maths lessons, the 9th grade teacher took his students to see, observe and take part in the whole process of flour production in a "Casa de Farinha", which is the place where manioc is processed and consists of a hut covered mostly with inaja straw, on a dirt floor, with no walls, where the oven and the other tools needed to process the manioc are located. It is usually located close to the fields and watercourses, but today it can also be located close to homes, due to the ease of using electricity owned by one of the community's residents (see figure 20a).

The activity began at five o'clock in the morning, the time when the cassava is harvested on the plantations. The students followed everything closely, recording in their notebooks notes on the information collected during the activity, information on production costs, the amount of cassava harvested, the total kilos of flour produced in each batch, the costs of transportation and packaging, the sale price and the profit made at the end.

According to the math teacher's report, the students themselves concluded that the final selling price of the flour, which is R$10.00 (ten reais) a kilo, is very cheap if you take into account all the work and physical effort involved, as a maximum of 20 kilos are obtained from each batch, and during the process in which the flour is roasted in the oven (a huge copper pot, round in shape, where the manioc is roasted to make the various types of flour) the workers take turns because they can't stop stirring, as the flour can burn or curdle.

The students also learned that after peeling the manioc, the producers grate it in a machine called a "catitu" (see figure 20c), which is a cylindrical piece of wood adorned with longitudinal steel serrations, used to grate the manioc, and that to extract the "tucupi" (juice extracted from the manioc, They use the "tipiti" (see figure 20b), which is a cylindrical, elongated object made from intertwined guaruma or jacitara splints, with elasticity, used to squeeze the manioc mass to remove the tucupi.

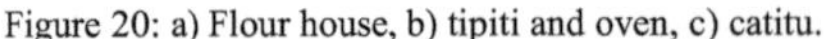
Figure 20: a) Flour house, b) tipiti and oven, c) catitu.

Source: Photographic collection of the authors.

At the end of the work, the students produced essays and questionnaires that generated tables and graphs about the whole process of producing and marketing flour, recording everything that was developed, understood and learned during this activity.

Knijnik et al. (2012, p. 18) say that

> Ethnomathematical thinking is centrally interested in examining practices outside of school, associated with rationalities that are not identical to the rationality that prevails in school mathematics [...] to look at these other rationalities is to think of other possibilities for Mathematics Education practiced in school.

During the research, we also followed the work of flour producers to observe how they use mathematics in their professional practice. These observations took place between January 2013 and March 2014. During this period, interviews were conducted with the workers and teachers at the Curiâu school.

During this period of observation, we looked at the mathematics content used in the day-to-day work of these professionals. First, we got to know the work of the flour producers in their daily lives, so that we could get closer to these workers. During the research, we carried out interviews, the instrument used in our data collection. At all times we observed that the residents of the community, the students, teachers and school staff were all very approachable and friendly and always welcomed us with great attention and availability, contributing significantly to our research work.

In Silva (2004), we find that the people of Curiaù have always been recognized as farmers, with a vocation for planting and raising, they have never stopped producing for their own sustenance and for commercialization, with the predominant crop being manioc. Curiaù is one of the few black communities in Brazil that, for a long time, has lived off planting and making manioc flour. To this day, the inhabitants produce manioc flour by hand to sell at the city fairs or within the community itself.

Manioc flour is an important product for the workers in the Curiaù community. During the research, we visited a flour mill located in the community to learn all about the process of making this food, which is a symbol of the subsistence of the flour producers who live in the community.

As we got to know a little more about the production of manioc flour, we discovered that for the residents of Curiaù this food is wrapped up in tradition and human development. We learned the stories of people who live off this flour production and that for these people it is much more than food, it is the means they have found to survive.

During the interviews we conducted, we identified the interviewer's questions by "P" and the interviewees' answers by -R".

One of the interviews begins with the interviewee singing:

I had a little bird stuck in its cage, it was midnight, it flew away. I was in my house, my judgment intervened, I only came to this meeting because the party-goer called me. (RAIMUNDA).

It was amidst the songs of one of Amapà's most traditional dances, the Marabaixo, that our interviewee arrived at the flour mill for another day's work. She told us that more than a thousand kilos of this food come out of Curiaù, which is one of the products most consumed by the people of Amapá, and that in order to reach the consumer's table there is a whole process, and it is from the hands of the workers of this community that the final product comes out.

In Mr. Eraldo's words, we learned that within the quilombola community of Curiaù there are three flour houses, which produce 30 batches, a weekly production of 600 kilos of flour, that is, each flour house produces an average of 10 batches and in the end each flour house is responsible for producing 200 kilos of flour.

During his interview, Mr. Eraldo explained to us the whole process of producing flour, from extracting the manioc to selling the product at the markets in Macapâ. We asked him what the producers do after the flour is ready:

Q: And once the flour is ready? What do you do with it? How do you package it?

A: The way to define quantities or volumes is in a practical way. You use a wooden box called a quart which has to contain 12 liters. A sack of flour is equivalent to 8 quarts, which was called 2 bushels of flour, because it contained 96 liters, which weighs 60 kilos. Decreases are used to define volumes: half a sack is equivalent to 4 quarts, which contains 48 liters, which is called 1 bushel, which weighs 30 kilos.

Q: And half a bushel?

A: Half a bushel is the equivalent of 2 quarts, which contains 24 liters, which weighs 15 kilos, and is called a pannier of flour.

Q: And a fourth?

A: A quarter is 12 liters, which weighs 7 and a half kilos.

Q: And does everyone here understand these divisions?

A: Today's consumer prefers to buy a can of flour, which contains a quarter and a half, which contains 18 liters. The difficulty is getting today's consumer to understand all these divisions." (ERALDO).

We can see from what Mr. Eraldo says that the mathematics used by Curiaù's flour producers can be taught to students in the classroom. In other excerpts from the interview with Mr. Eraldo, we learned that the Curiaù flour producers work 13-hour days, starting at 5am and finishing at 6pm. They also use the palm and finger as measurements when cutting the manioc stalk for planting, which is equivalent to 35 cm. The size of the spacing they use is 50 cm from one foot to the next, and the amount of weeding they do before harvesting, which takes place a year after planting. Mrs. Raimunda tells us about the time it takes for the batch to be ready, on average two and a half hours.

Going back to Mr. Eraldo's interview, we can also see how long it takes to toast the flour, which varies between 35 and 60 minutes. The use of the -quarta", a wooden box, which he explained to us has the capacity to store 12 liters of flour. We noted the ease with which he explained to us that 1 sack of flour contains 8 quartas, which is equivalent to 2 bushels, which is equivalent to 96 liters, which corresponds to 60 kilos of flour produced. By using the term "decreases", he makes the following mathematical connections:

- 1/2 shakeflour =4quarts = 48 liters = 1 bushel = 30 kg of flour
- 1/4 flour shaker =2quarts = 24 liters = 1/2 bushel = 15 kg of flour
- 1/8 flour shaker =1quart = 12 liters = 1/4 bushel = 7.5 kg of flour

When he tells us about consumers' preference for buying 1 can of flour, he shows us that 1 can of flour is equivalent to 1 and a half quarts, which corresponds to 18 liters of flour. The sale value of a kilo of flour, which is R$10.00, the total production of 600 kilos of flour, the profit made on sales, production costs, packaging costs, transportation costs, fuel costs - all this information can be explored in the classroom by math teachers.

During the research, we also realized that the flour producers in the quilombola community of Curiù are not aware of the work of the Brazilian Agricultural Research Corporation (EMBRAPA) in Amapà. In Embrapa (2011), we found that the work of EMBRAPA/AP is to help flour production in the state, committed to rural extension and to drawing up projects that help develop flour production. The institution is working on a project that will help boost cassava flour production in the state of Amapà. This project is still being finalized, and EMBRAPA/AP researchers are testing local varieties and those from other states as well, in the cities of Macapâ, Manaus, Belém and fruit-growing cassava in the state of Bahia. Among these varieties, EMBRAPA will be able to indicate to flour producers five varieties of cassava that are much more productive than those that are currently being cultivated on Amapà's land. According to EMBRAPA/AP researchers, the state's production is 12 tons per hectare of cassava, but they say that with these varieties that are being tested, production could reach 30 and even 40 tons of cassava. This project will be presented by EMBRAPA/AP to Amapà's city halls and to farmers interested in learning about and applying this production system using these other cassava varieties.

Also in Embrapa (2011), we find that the Rural Development Institute of Amapà - RURAP, works with rural extension and technical assistance. This body has been developing a Territorial Program in Family Farming - PROTAF, which is one of Amapà's public policy programs. PROTAF provides family farmers with one hectare per family, per family unit, helping them to work with the soil, helping to correct and fertilize it to receive crops, where the main crop is cassava. During the research, we found that Curiaù's flour producers are unaware of RURAP's rural extension and technical assistance work, and that they could look for this body to receive this support. In 2013, RURAP assisted 15,000 family farmers, but they estimate that there are more than 25,000 farmers in the whole of the state of Amapà, and the goal of RURAP's researchers is to assist all these family farmers, because they claim that data from the Brazilian Institute of Geography and Statistics - IBGE, shows that family farmers assisted by rural extension have an income up to four times higher than those who are not assisted.

PROTAF researchers say that Amapà's climate and soil are favorable for growing cassava. The EMBRAPA researchers say that cassava is a rustic plant, which adapts to all climates and all types of soil, especially in hot and humid regions. With plenty of water and plenty of heat, cassava adapts well and consequently produces well. But these researchers explain that the soil in Amapá is weak, so it is necessary to work on fertilization, and this is precisely the work of these researchers, which is to provide farmers with a fertilization system that improves production, enabling cassava to develop its full productive potential in the field.

According to EMBRAPA (2001), studies show that the average Amapá resident consumes 35 kilos of manioc flour every year. In 2000 alone, more than 150,000 tons of manioc were produced, and even so, this amount failed to supply the local market. And although this production has increased a lot, according to EMBRAPA researchers, it's still not enough, because consumption has increased a lot, which is why PROTAF came to optimize flour production, but it hasn't succeeded yet. With the completion of the project, EMBRAPA/AP believes that at least self-sufficiency in cassava flour production in Amapà will be possible.

FINAL CONSIDERATIONS

We've seen that the Curiaù Community school seeks to develop work that involves the community, producing knowledge that seeks applicability in its form of knowledge established in partnership with the workers who produce and sell the flour, tucupi, açai and other products they sell, making a substantial contribution to the processes of teaching and learning mathematics.

We can see that the teachers at this school are aware of the importance of the contributions emerging from the culture of people of African origin, in accordance with what is proposed in Federal Law 10.639/03. The curriculum guidelines emanating from this law make it possible for Mathematics teachers to adopt historical and intercultural approaches, which broaden the focus of the Brazilian school curriculum towards cultural diversity, marked by an African origin, whose roots lie in the colonial period that produced ethnic and cultural heritages.

We also realized that we can relate the guidelines of Law 10.639/03 to Ethnomathematics. This relationship allows us to believe that the compulsory inclusion of African and Afro-Brazilian history and culture in school curricula may in fact become a reality, contributing substantially to the teaching and learning of mathematics.

From the outset, the research provided us with moments of reflection, criticism and suggestions for improving our practice and educational practice in general. By observing the work activities of Curiaù's flour producers, we were able to ascertain which mathematical contents are most used by these workers.

This work has made it clear that Ethnomathematics can guide the production of knowledge in the world, in favor of the development of an increasingly egalitarian society, devoid of racial or cultural prejudices, seeking the formation of citizens aware of the ethnic issues that shape the Brazilian educational scenario, aware of the importance of the contributions emerging from the culture of peoples of African origin, in accordance with what is proposed in Brazilian Federal Law 10.639/03.

We believe that holding the workshop at the Curiâu school was important for understanding the educational process at this elementary school, and the possibility of carrying out an activity using a game of African origin was significant. According to the students' responses and teacher Dalva's report, we believe that the workshop made a positive and significant contribution to the teaching and learning processes in the classroom.

We believe that Law 10.639/03 can be discussed in mathematics classes, in schools and universities, highlighting Afro-Brazilian civilizational values and knowledge of African origin, according to Vergani (2000) who emphasizes the importance of an Ethnomathematical education that works with -the rational, psychic, emotional, social and cultural wholeness of man", assuming -a creative posture that echoes different levels and different degrees of depth" (VERGANI. 2000, p.41), overcoming the

It is therefore possible to think of an ethnomathematics education for ethnic-racial relations in favor of valuing African culture and sciences. In this way, it is possible to think of an Ethnomathematical Education for ethnic-racial relations in favor of valuing the culture and sciences of African origin, acting on discrimination and exclusion and seeking the full exercise of citizenship.

The educational and challenging approach of reflecting on the importance of implementing Law 10.639/03 as an essential reinforcement in the construction of a policy of citizen training, based on Ethnomathematics, so that we can meet the aspirations of education on ethnic-racial relations, as well as the inclusion of the theme of Afro-Brazilian and African Culture, we emphasize that it is necessary to discuss and reflect on the improvement of educational quality and to emphasize our commitment to promoting equality.

We can conclude that this work is a reflection of a discussion that is being consolidated on the educational scene, as we believe that the effective implementation of Federal Law 10.639/03 will contribute substantially to the policy of comprehensive education and stimulate debate and awareness of the social commitments of education. We hope that this book can contribute to other researchers in Mathematics Education, so that more research can be carried out on this subject and encourage or awaken in other teachers the search for or interest in this knowledge or the production of more materials of the same or similar nature.

REFERENCES

ABRIL, Almanac. **Society**: Quilombo. Sao Paulo: April 2013.

AJAYI, Jacob Festus Adeniyi. **General History of Africa**: Africa from the 19th century to the 1880s. - Vol. VI. 2ª ed. - Sao Paulo: Cortez, 2011.

AMAPA. Decree No. 1417, of September 28, 1992. With the aim of protecting and conserving natural and environmental resources, creates the Curiaù Environmental Preservation Area, in the Municipality of Macapà. **Official Gazette of the State of Amapà**, Macapà, AP, September 28, 1992.

AMAPA. Decree No. 0197, of January 23, 2001. Officializes the creation and naming of the José Bonifàcio State School, in the Municipality of Macapà. **Official Gazette of the State of Amapà**, Macapà, AP, January 24, 2001.

ARANHA, Maria Lùcia de Arruda; MARTINS, Maria Helena Pires. **Filosofando**: Introduction to Philosophy. - 4th ed. - Sao Paulo: Moderna, 2009.

BAENA, Antônio Ladislau Monteiro. **Compendia das Eras da Provincia do Parà**. Belém: UFPA, 1969.

BRAZIL. **National Curriculum Parameters**: Mathematics. Secretariat for Primary Education. Brasilia: MEC/SEF, 1997.

BRASIL. **Parâmetros Curriculares Nacionais do Ensino Mèdio**: Matemàtica. Secretariat for Basic Education. Brasilia: MEC/SEB, 1999.

BRAZIL. Law No. 10.639, of January 9, 2003. Amends Law No. 9.394, of December 20, 1996, which establishes the guidelines and bases of national education, to include the subject of "Afro-Brazilian History and Culture" in the official school curriculum, and makes other provisions. **Diàrio Oficial da Uniao**, Brasilia, DF: MEC, Jan. 10, 2003.

BRASIL. **Diretrizes Curriculares Nacionais para a Educaçao das Relações Étnico-raciais e para o Ensino de História e Cultura Afro-Brasileira e Africana**. Brasilia: MEC, 2004.

BOYER, Carl Benjamin. **História da Matemàtica** (Translation of: A history of mathematics, by: Elza F. Gomide). 2 ed. Sao Paulo: Blücher, 1996.

BROUGÈRE, Gilles. **Play and education**. Porto Alegre: Artes Médicas, 1998.

CHAGAS, Marco Antônio. **Curiau:** Environmental Preservation Area Dossier. Macapà: SEMA, 1997.

CONAQ/AP - **State Coordination of Articulation of Black Rural Quilombola Communities of Amapà**, 2014. Available at http://quilombolasdoamapa.blogspot.com.br/. Accessed on 05/02/2017.

COSTA, Wanderleya. Indigenous and Afro-Brazilian histories and cultures in math classes. **Educaçao em Revista**. Belo Horizonte: UFMG, v. 25, p. 175-197, 2009.

COSTA, Wanderleya; OLIVEIRA, Cristine. Mathematics education and racial prejudice: African and Afro-Brazilian cultures in the classroom. In: X ENCONTRO NACIONAL DE EDUCAÇÀO MATEMATICA, 2010, Salvador. **Proceedings**... Salvador: SBEM, 2010. 1 CDROM.

CUNHA JR., Henrique. Afroethnomatics. **Revista Temas em Educaçao**. Vol.3, p. 83-95, [S.l], 2004.

D'AMBROSIO, Ubiratan. **Ethnomathematics**: the link between tradition and modernity. Belo Horizonte: Autèntica, 2011.

EMBRAPA. **Manual of Good Agricultural Practices and APPC System**. Food Quality and Safety Series. Brasilia: EMBRAPA/SEDE, 2001.

EMBRAPA. **Planning a cassava flour mill**. Food Quality and Safety Series. Macapà: EMBRAPA/MACAPA/AP, 2011.

EVES, Howard. **Introduction to the History of Mathematics** (Translation of: An introduction to the history of mathematics, by: Higyno H. Domingues). Campinas: Unicamp, 2004.

FACUNDES, Francinete da Silva; GIBSON, Valdeci Marques. **Natural resources and environmental diagnosis of the Curiaù River APA**. 58 f. Course Conclusion Paper (Graduation in Geography) - Federal University of Amapà, Macapà, 2000.

FRANÇA, Jean Marcel Carvalho; FERREIRA, Ricardo Alexandre. **Três vezes Zumbi:** a Construçào de um herói brasileiro. Sào Paulo: Très Estrelas, 2012.

FRANKENSTEIN, Marilyn; POWELL, Arthur. **Ethnomathematics:** challenging eurocentrism in Mathematics education. Albany: State University of New York Press, 1997.

GERDES, Paulus. **From ethnomathematics to art-design and cyclic matrices**. Belo Horizonte: Autèntica, 2010.

GIORDANI, Màrio Curtis. **History of Africa**: before the discoveries. Petrópolis: Vozes, 2013.

KNIJNIK, Gelsa. **Exclusion and resistance:** mathematics education and cultural legitimacy. Porto Alegre: Artes Médicas, 1996.

KNIJNIK, Gelsa; WANDERER, Fernanda; GIONGO, Ieda Maria; DUARTE, Claudia Glavam. **Ethnomathematics in movement**. Belo Horizonte: Autèntica, 2012.

LIMA, Heloisa Pires; GNEKA, Geórge; LEMOS, Màrio. **The seed that came from Africa.** Sào Paulo: Salamandra, 2005.

MACIEL, Alexsara de Souza. **Conversa amarra preto**: a trajectória histórica da uniao dos negros do Amapà. 181 f. Master's dissertation, Postgraduate Program, State University of Campinas, Campinas, 2001.

MARIN, Rosa Elisabeth Acevedo. **Born in Curiàu**: identification report presented to the Palmares Cultural Foundation. Belém: NAEA/UFPA, 1997.

MATTOS, José Roberto Linhares de; BRITO, Maria Leopoldina Bezerra. Rural agents and their professional practices: a link between mathematics and ethnomathematics. **Ciência & Educaçao**, Bauru, vol. 18, n.4, p. 965-980, 2012.

MAZRUI, Ali Alamin; WONDJI, Christophe. **General History of Africa**: Africa since 1935. - Vol. VIII. 2ª ed. - Sao Paulo: Cortez, 2011.

MENDES, Iran Abreu. **Mathematics and investigation in the classroom:** weaving cognitive networks in learning. Sao Paulo: Livraria da Fisica, 2009a.

MENDES, Iran Abreu. **Historical Investigation in the Teaching of Mathematics**. Rio de Janeiro: Ciência Moderna Publishing House, 2009b.

MONTEIRO, Alexandrina. Some reflections on the educational perspective of Ethnomathematics. **Zetetiké**, v.12, n.22, Campinas, p. 9-32, 2004.

MORAIS, Paulo Dias. **History of Amapà**: the past is the mirror of the present. Macapâ: JM, 2009.

MORAIS, Paulo Dias. **Amapà in Perspectives:** municipalities of Amapà. Macapâ: JM, 2011.

MUNIZ, Cristiano Alberto. **Jeux de société et activité mathématique chez l'enfant.** 629 f. Thesis (Doctorate in Educational Sciences) - Postgraduate Program in Educational Sciences, Université Paris Nord, France, 1999.

OLIC, Nelson Bacic; CANEPA, Beatriz. **Africa**: land, societies and conflicts. Sao Paulo: Moderna, 2012.

OLIVEIRA, Cristiane Coppe. The ethnomathematics program and the ethnic-racial context in teaching practice. In: XIII INTERAMERICAN CONFERENCE ON MATHEMATICAL EDUCATION, 2011, Recife. **Proceedings**... Recife: EDUMATEC-UFPE, 2011. 1 CD-ROM.

PENNAFORTE, Charles. **Africa:** horizons and challenges for the 21st century. Sao Paulo: Atual, 2009.

QUEIROZ, Silvaneide**. Curiaù Quilombola Territory and Curiaù River Environmental Preservation Area**: interpretations of socio-environmental conflicts through ecological economics. 2007. Master's dissertation from the Postgraduate Program at the Federal University of Parà, Belém, 2007.

RÊGO, Rogéria Galdêncio; RÊGO, Rômulo Marinho. **Matematicativa**. Joao Pessoa: University, 2000.

RIPOLL, Oriol; CURTO, Rosa Maria. **Games for the whole world.** Sao Paulo: Ciranda cultural, [s/d].

SANTOS, Celso José dos. **African Games and Mathematics Education:** sowing seeds with the Mancala family. Master's dissertation, Maringà State University, Maringà, 2008.

SANTOS, Lorene. **Teaching African and Afro-Brazilian history and culture**: dilemmas and challenges of the reception to Law 10.639/03. Rio de Janeiro: Pallas, 2013.

SANTOS, Maximina Magda de França. Continuing teacher training from the perspective of ethnomathematics based on African cultures: advances and obstacles. In: XIII INTERAMERICAN CONFERENCE ON MATHEMATICAL EDUCATION, 2011, Recife.

Anais... Recife: EDUMATEC-UFPE, 2011. 1 CD-ROM.

SILVA, Sebastiao Menezes da. **Curiaù:** his life, his story. Macapâ: Valcan, [n.d.].

SILVA, Sebastiao Menezes da. **Curiaù:** its life, its history. Macapâ: FUNDECAP, 2000.

SILVA, Sebastiao Menezes da. **Curiaù:** the resistance of a people. Macapâ: SEMA, 2004.

SILVA, Tomaz Tadeu da. **Documents of Identity:** an introduction to curriculum theories. Belo Horizonte: Autèntica, 1999.

SOARES, Marcelo André; RODRIGUES, Maria Emilia Brito. **Amapà:** living our history. Curitiba: Base, 2008.

SOUZA, M. M. **The game and mathematical learning in higher education.**

Master's dissertation, University of Brasilia, Brasilia, 2003.

SUDAM. **Climatological Atlas of the Brazilian Amazon**. Belém: Sudam, 1984.

TRIGO, Ilda. **Transdisciplinary Africa**: rediscovering Brazil. In: Educatrix, n.4, Sao Paulo: Moderna, 2013.

TRINDADE, Azoilda Loretto da. **In search of full citizenship**. In: Saberes e fazeres, v.1, Rio de Janeiro: Roberto Marinho Foundation, 2006.

TRINDADE, Azoilda Loretto da. **Dialoguing with challenges**. In: Implementation of the Curricular Guidelines for Ethnic-Racial Relations Education and the Teaching of Afro-Brazilian and African History and Culture in Professional and Technological Education. Brasilia: MEC/SETEC, 2008.

VERGANI, Teresa. **Ethnomathematical Education:** What is it? Lisbon: Pandora Editions, 2000.

VIDEIRA, Piedade Lino. **Marabaixo, afro-descendant dance:** signifying the ethnic identity of black people from Amapá. Fortaleza: UFC, 2009.

VIDEIRA, Piedade Lino. **Batuques, folias e ladainhas:** the culture of the Cria-ύ quilombo in Macapâ and its education. Fortaleza: UFC, 2013.

VYGOTSKY, Lev Semenovich. **The social formation of the mind**. Sao Paulo: Martins Fontes, 1991.

VISENTINI, Paulo Fagundes; RIBEIRO, Luiz Dario Teixeira; PEREIRA, Analùcia Danilevicz. **History of Africa and Africans**. Petrópolis: Vozes, 2013.

ZASLAVSKY, Claudia. **Mathematical games and activities for the whole world:** multicultural fun for ages 8 to 12. Porto Alegre: Artmed, 2000.

Website: http://casteloroger.blogspot.com.br/2011/06/areas-protegidas-quadro-geral-no-amapa.html. Accessed on: 18/05/2013.

Website: http://www.baixarmapas.com.br/mapa-da-africa. Accessed on: 29/11/2013.

Website: http://menrvatemplodosaber.blogspot.com.br/2014_01_01_archive.html. Accessed on: 17/02/2014.

VIEIRA, Pedro Lino. Mandombe: afro-descendant dances [illegible] black people from Angola. [illegible], 2009.

[illegible] Pedro Lino. [illegible] culture [illegible] Ed. UFC, 2015.

[illegible] The social [illegible] of the [illegible]

[illegible] of Africa [illegible]

[illegible]

[illegible]

Printed by Books on Demand GmbH, Norderstedt / Germany